湖北省化学工程与工艺专业校企合作联盟系列教材

Physical Chemistry Experiment

物理化学实验

（双语）

- 刘安昌 主编　　- 田琦峰　赵慧平 副主编

化学工业出版社

·北京·

内容简介

本书在多年实验教学改革与实践基础上编写而成。全书共分为 5 章,在阐述物理化学实验的目的要求、实验的安全防护、测量误差及分析、实验数据的表达方法的基础上,重点介绍了热力学、电化学、化学动力学及界面化学等相关的 14 个基础实验,以及含综合及设计性的提高型实验共 15 个项目,另外还简单介绍了实验技术与仪器、常用数据表。

本书可作为高等学校化学、化工、应用化学、材料科学、环境科学、生命科学等相关专业的物理化学实验教材,也可供高校教师及从事化学研究的技术人员参考。

图书在版编目(CIP)数据

物理化学实验(双语)/刘安昌主编.—北京:化学工业出版社,2012.12(2024.8重印)
湖北省化学工程与工艺专业校企合作联盟系列教材
ISBN 978-7-122-15563-4

Ⅰ.①物⋯ Ⅱ.①刘⋯ Ⅲ.①物理化学-化学实验-高等学校-教材-汉、英 Ⅳ.①O64-33

中国版本图书馆 CIP 数据核字(2012)第 244232 号

| 责任编辑:杜进祥 | 文字编辑:向 东 |
| 责任校对:吴 静 | 装帧设计:韩 飞 |

出版发行:化学工业出版社(北京市东城区青年湖南街 13 号 邮政编码 100011)
印　　装:北京七彩京通数码快印有限公司
787mm×1092mm 1/16 印张 11¼ 字数 282 千字 2024 年 8 月北京第 1 版第 8 次印刷

购书咨询:010-64518888　　　　　　　　售后服务:010-64518899
网　　址:http://www.cip.com.cn
凡购买本书,如有缺损质量问题,本社销售中心负责调换。

定　价:30.00元　　　　　　　　　　　　　　　版权所有　违者必究

序

高等教育竞争的本质是人才培养质量的竞争。《国家中长期教育改革和发展规划纲要（2010—2020年）》对高等教育的未来发展提出了明确要求，即全面提高高等教育质量、提高人才培养质量、提升科学研究水平、增强社会服务能力、优化结构办出特色，力争到2020年，高等教育结构更加合理，特色更加鲜明，人才培养、科学研究和社会服务整体水平全面提升。为此，教育部发布了《关于实施卓越工程师教育培养计划的若干意见》（教高〔2011〕1号）以及《关于"十二五"期间实施"高等学校本科教学质量与教学改革工程"的意见》（教高〔2011〕6号），其目的是进一步深化本科教育教学改革，提高本科教育教学质量，大力提升人才培养水平和创新能力。

武汉工程大学（原武汉化工学院）建校于1972年，经过40年的发展，办学特色日趋鲜明：化工及相关学科为主、多学科协调发展的学科专业特色；适应复合应用型、创新型人才培养目标要求的工程教育特色；"立足湖北、辐射全国，服务行业和区域经济社会发展"的服务面向特色。学校紧密围绕人才培养、科学研究、服务社会、传承文化的主题，确立了"质量立校、科技强校、人才兴校、突出特色、协调发展"的办学思路以及"以质量为根本，以网络为基础，以开放为特点，以创新为动力"的教学指导思想。作为一所有着较强行业背景的地方高校，积极参与实施协同创新和卓越工程师教育培养计划，既是推动教育与科技、经济、文化紧密结合，建设创新型国家的战略行动，也是提高学校核心竞争力、服务行业和区域经济发展、实现学校新跨越的重要过程。

为了充分发挥高等学校的教学科研优势，加快建设以企业为主体、市场为导向、产学研相结合的技术创新体系，探索化工类人才培养的改革创新之路，服务地方经济建设，按照"学科引领、合作发展、共建共享、彰显特色、服务地方"的指导思想，2011年7月，由武汉工程大学倡议发起的"湖北省化学工程与工艺专业校企合作联盟成立大会"在武汉顺利召开，加入联盟的有华中科技大学、武汉理工大学、湖北大学、长江大学、三峡大学等20余所高校以及中石化武汉分公司、武汉有机实业有限公司、湖北祥云（集团）化工股份有限公司等10余家企业。"湖北省化学工程与工艺专业校企合作联盟"的成立不仅有利于加强湖北省内各高校之间化学工程与工艺专业之间的联系，有效实现资源共享，而且有利于促进高等学校和企业之间的交流与合作，共同探讨新形势下如何提高化工类专业人才的培养质量和针对性，为化学工业的发展培养优秀的工程技术人才，进而推动化工行业和区域经济发展。

在国家建设资源节约型、环境友好型社会的大背景下，石油化工、矿产资源等领域发展空间巨大，化学工业发展将是国家新型工业化的战略重点，化学工业也是湖北省国民经济的支柱产业之一。在 40 年的办学历程中，学校始终注重学生工程实践能力和创新能力的培养，注重教学与科研和生产实际相结合，逐步构建了以实训-实验-实习-创新为主要内容的"三实一创"实践教学体系。本次由化学工业出版社出版的《环境与化工清洁生产创新实验教程》、《化工原理课程设计》、《化工设计》、《化学工程与工艺实习指南》、《化工工艺学》、《化工原理实验（双语）》、《物理化学实验（双语）》等系列教材汇集了"湖北省化学工程与工艺专业校企合作联盟"部分高校和企业的教学科研开发成果，旨在紧跟化工行业发展前沿和社会需求，适时调整人才培养计划，更新教学内容和教学方法，创新课程体系，深化教学改革，为逐步形成专业发展与社会需求相适应的人才培养体系添砖加瓦。

谨此为序。

吴元欣

2012 年 7 月于武汉工程大学

前言

本书是在邝生鲁、刘常坤、向建敏编写的《物理化学实验》（原武汉工业大学出版社，1996年）的基础上，经修改补充后完成的。为适应21世纪化工类人才培养对实验教学的要求以及创建国家级化学化工实验教学示范中心的要求，我们将物理化学实验划分为基础实验和提高型实验（含综合与设计实验）两个层次，修改、调整和增补了部分基础实验，设计性实验和大多数综合实验为新增内容。其中基础实验为双语教学内容。

全书分为五章。

(1) 绪论：介绍物理化学实验的目的要求、实验的安全防护、测量误差及分析、实验数据的表达方法。

(2) 基础实验：包括热力学、电化学、化学动力学及界面化学等内容共14个实验。

(3) 提高型实验：含综合及设计性实验共15个。教学实践表明，提高型实验有利于提高学生的学习兴趣和解决实际问题的能力。

(4) 实验技术与仪器。

(5) 常用数据表。

本书的编写吸收了教研室教师多年来的科研成果，是本课程教师长期从事物理化学实验教学工作的共同结晶。本书由刘安昌主编，田琦峰、赵慧平副主编，刘安昌统稿。参加编写的老师有贾丽慧、向建敏、孙雯、丁莉莉、袁军、朱想明、黄华良和黄飞。孙炜对本书的编写提供了帮助和指导，在此谨致谢忱。

由于编者水平有限，书中不足之处在所难免，恳请读者批评指正。

编者

2014年1月

目 录

第 1 章　绪论 ⋯⋯⋯⋯⋯⋯⋯⋯⋯⋯⋯⋯⋯⋯⋯⋯⋯⋯⋯⋯⋯⋯⋯⋯⋯⋯⋯⋯⋯⋯⋯⋯⋯⋯⋯ 1
　1.1　物理化学实验的目的和要求 ⋯⋯⋯⋯⋯⋯⋯⋯⋯⋯⋯⋯⋯⋯⋯⋯⋯⋯⋯⋯⋯⋯⋯⋯⋯ 1
　1.2　物理化学实验中的安全防护 ⋯⋯⋯⋯⋯⋯⋯⋯⋯⋯⋯⋯⋯⋯⋯⋯⋯⋯⋯⋯⋯⋯⋯⋯⋯ 2
　1.3　测量误差及分析 ⋯⋯⋯⋯⋯⋯⋯⋯⋯⋯⋯⋯⋯⋯⋯⋯⋯⋯⋯⋯⋯⋯⋯⋯⋯⋯⋯⋯⋯⋯ 5
　1.4　实验数据的表达方法 ⋯⋯⋯⋯⋯⋯⋯⋯⋯⋯⋯⋯⋯⋯⋯⋯⋯⋯⋯⋯⋯⋯⋯⋯⋯⋯⋯ 16

第 2 章　基础实验 ⋯⋯⋯⋯⋯⋯⋯⋯⋯⋯⋯⋯⋯⋯⋯⋯⋯⋯⋯⋯⋯⋯⋯⋯⋯⋯⋯⋯⋯⋯⋯ 22
　实验 2.1　恒温槽的性能调节及液体黏度测定 ⋯⋯⋯⋯⋯⋯⋯⋯⋯⋯⋯⋯⋯⋯⋯⋯⋯⋯ 22
　Experiment 2.1　Performance of Thermostatic Bath and Determination
　　　　　　　　　of Liquid Viscosity ⋯⋯⋯⋯⋯⋯⋯⋯⋯⋯⋯⋯⋯⋯⋯⋯⋯⋯⋯⋯⋯ 25
　实验 2.2　易挥发物质摩尔质量的测定 ⋯⋯⋯⋯⋯⋯⋯⋯⋯⋯⋯⋯⋯⋯⋯⋯⋯⋯⋯⋯⋯ 29
　Experiment 2.2　Determination of Molar Mass by Evaporization ⋯⋯⋯⋯⋯⋯⋯⋯⋯⋯ 31
　实验 2.3　溶解热的测定 ⋯⋯⋯⋯⋯⋯⋯⋯⋯⋯⋯⋯⋯⋯⋯⋯⋯⋯⋯⋯⋯⋯⋯⋯⋯⋯⋯ 33
　Experiment 2.3　Determination of Heat of Solution ⋯⋯⋯⋯⋯⋯⋯⋯⋯⋯⋯⋯⋯⋯⋯⋯ 39
　实验 2.4　燃烧热的测定 ⋯⋯⋯⋯⋯⋯⋯⋯⋯⋯⋯⋯⋯⋯⋯⋯⋯⋯⋯⋯⋯⋯⋯⋯⋯⋯⋯ 43
　Experiment 2.4　Determination of Heat of Combustion ⋯⋯⋯⋯⋯⋯⋯⋯⋯⋯⋯⋯⋯⋯ 45
　实验 2.5　液体饱和蒸气压的测定 ⋯⋯⋯⋯⋯⋯⋯⋯⋯⋯⋯⋯⋯⋯⋯⋯⋯⋯⋯⋯⋯⋯⋯ 49
　Experiment 2.5　Determination of Saturated Vapor Pressure of a Pure Liquid ⋯⋯⋯⋯⋯ 52
　实验 2.6　二组分汽-液平衡相图的测定 ⋯⋯⋯⋯⋯⋯⋯⋯⋯⋯⋯⋯⋯⋯⋯⋯⋯⋯⋯⋯⋯ 54
　Experiment 2.6　Determination of Phase Diagram of a Two-component Liquid-vapor
　　　　　　　　　Equilibrium System ⋯⋯⋯⋯⋯⋯⋯⋯⋯⋯⋯⋯⋯⋯⋯⋯⋯⋯⋯⋯⋯ 57
　实验 2.7　电导率的测定及应用 ⋯⋯⋯⋯⋯⋯⋯⋯⋯⋯⋯⋯⋯⋯⋯⋯⋯⋯⋯⋯⋯⋯⋯⋯ 60
　Experiment 2.7　Determination and Applications of Electrical Conductance ⋯⋯⋯⋯⋯⋯ 63
　实验 2.8　电动势的测定及其应用 ⋯⋯⋯⋯⋯⋯⋯⋯⋯⋯⋯⋯⋯⋯⋯⋯⋯⋯⋯⋯⋯⋯⋯ 67
　Experiment 2.8　Determination of Electromotive Force of Reversible Cell ⋯⋯⋯⋯⋯⋯⋯ 71
　实验 2.9　蔗糖水解速率常数的测定 ⋯⋯⋯⋯⋯⋯⋯⋯⋯⋯⋯⋯⋯⋯⋯⋯⋯⋯⋯⋯⋯⋯ 76
　Experiment 2.9　Determination of Rate Constant for Hydrolysis of Sucrose ⋯⋯⋯⋯⋯⋯ 79
　实验 2.10　过氧化氢分解速率常数的测定 ⋯⋯⋯⋯⋯⋯⋯⋯⋯⋯⋯⋯⋯⋯⋯⋯⋯⋯⋯⋯ 82
　Experiment 2.10　Catalytic Decomposition of Hydrogen Peroxide ⋯⋯⋯⋯⋯⋯⋯⋯⋯⋯ 85
　实验 2.11　乙酸乙酯皂化反应动力学参数的测定 ⋯⋯⋯⋯⋯⋯⋯⋯⋯⋯⋯⋯⋯⋯⋯⋯ 88

Experiment 2.11　Determination of Kinetic Parameters for Saponification of
　　　　　　　　Ethyl Acetate ·· 90
实验 2.12　溶液表面张力的测定 ·· 93
Experiment 2.12　Determination of Surface Tension of Solutions ············ 96
实验 2.13　凝固点降低法测定溶质的摩尔质量 ·································· 99
Experiment 2.13　Determination of Solute Molar Mass by Freezing-point-
　　　　　　　　depression Method ·· 102
实验 2.14　二组分金属相图的测定 ·· 106
Experiment 2.14　Solid-Liquid Binary Phase Diagram ························ 109

第 3 章　提高型实验 ·· **112**
实验 3.1　金属电极的制作、封装及电位测定 ··································· 112
实验 3.2　洗手液的研制及性能测定 ··· 113
实验 3.3　氨基甲酸铵分解热力学函数的测定 ·································· 115
实验 3.4　溶胶的制备及 ζ 电势的测定 ·· 118
实验 3.5　沉降分析 ·· 120
实验 3.6　接触角的测定 ··· 123
实验 3.7　B-Z 化学振荡反应活化能的测定 ······································ 125
实验 3.8　差热及热重分析 ·· 127
实验 3.9　金属腐蚀行为的电化学研究 ·· 131
实验 3.10　粉末润湿性能的测定 ·· 136
实验 3.11　活性炭比表面积的测定 ··· 137
实验 3.12　油品燃烧热的测定 ·· 140
实验 3.13　微溶盐浓度积的测定 ·· 141
实验 3.14　极化曲线的测定 ··· 142
实验 3.15　不同反应体系化学振荡现象的初步研究 ··························· 144

第 4 章　实验技术与仪器 ·· **147**
4.1　大气压力计 ··· 147
4.2　温度控制系统 ··· 149
4.3　饱和蒸气压减压装置 ··· 152
4.4　阿贝折光仪 ··· 153
4.5　精密数字压力计 ··· 156
4.6　数字式贝克曼温度计 ··· 157
4.7　DDS-11A 型电导率仪 ·· 158
4.8　补偿法原理及 UJ-25 型高电势直流电位差计 ··························· 159
4.9　旋光仪的构造原理及使用方法 ·· 162

4.10	WLS-2 可调式恒流电源	164
4.11	SWC-ⅡD 精密数字温度温差仪	165

附录　物理化学常用数据表 167

附表 1	国际相对原子质量	167
附表 2	水的饱和蒸气压	168
附表 3	水的表面张力	169
附表 4	水的黏度	169
附表 5	甘油水溶液黏度	169
附表 6	几种液体的饱和蒸气压	170
附表 7	醋酸的标准电离平衡常数	170
附表 8	不同温度下氯化钾的溶解热	170
附表 9	KCl 水溶液的电导率	170
附表 10	压力单位换算	171
附表 11	能量单位换算	171
附表 12	气瓶颜色标志一览	171

参考文献 172

第1章 绪 论

化学是建立在实验基础上的科学。物理化学实验是化学实验课程的一个重要分支。它综合了化学领域中各分支所需的基本实验工具和研究方法,主要是应用物理学原理与技术,使用仪器或若干仪器组合成的测量体系,对系统的某一物理化学性质进行测量,进而研究化学问题。其研究方法和实验技能是化学工作者必须具备的基本功。因而是化学、化工类以及与之关系密切的多个学科专业学生必修的一门重要基础实验课程。

1.1 物理化学实验的目的和要求

1.1.1 物理化学实验的目的

① 使学生了解物理化学的实验方法,掌握物理化学的基本实验技术和技能,学会测定物质特性的基本方法,熟悉物理化学实验现象的观察与记录、实验条件的判断与选择、实验数据的测量与处理、实验结果的分析与归纳等一套严谨的实验方法,从而加深对物理化学基本理论和概念的理解。

② 通过实验培养学生的实验能力、创新思维能力与进行初步科研的能力。首先,学生在实验中通过思考、分析、对比、综合归纳才能得出实验结果,这个过程培养了学生的逻辑思维能力和创造力。其次,物理化学实验不同于其他的基础化学实验,它是由学生自己预习教材、参考书、仪器使用说明书和工具书等,自己完成实验任务,教师只是指导实验,而不是给学生讲实验。因此,学生独立完成物理化学实验能极大地锻炼学生的自学能力。此外,通过综合与设计性实验的训练,学生应能够根据某一具体的目的要求,查阅资料、根据实验原理拟定实验方案、选用合适的配套仪器、设计实验步骤、测定和处理测量数据,提高进行一般实验研究工作的能力(如毕业论文、实验设计等)。

③ 培养学生观察实验现象,正确记录和处理数据,进行实验结果的分析和归纳,以及书写规范、完整的实验报告等能力,并养成严肃认真、实事求是的科学态度和作风。

1.1.2 物理化学实验的基本要求

(1) 预习 在进行实验操作之前,必须仔细阅读实验教材及有关参考资料,明确实验目的要求,弄懂实验原理和方法,了解实验中所使用的仪器以及具体操作步骤,在此基础之上,写出预习报告(内容包括简要实验原理和实验方法,实验操作步骤,注意事项,数据记录表格,以及预习中出现的疑难点等)。预习报告写在专用的实验记录本上,每次进实验室都要携带实验记录本，经指导教师检查预习报告后才能进行实验操作。

对于设计性实验，应按实验项目要求拟定好实验方案（其他要求同上）。

实践证明，实验前的预习是否充分，不仅直接影响实验效果，而且关系到实验能否正常进行。

（2）实验操作

① 待指导教师讲述实验要点和注意事项后（通常是学生在预习时不易解决的问题），检查仪器药品是否合乎实验要求，记录实验条件（室温、大气压、药品纯度、仪器精度等），做好各项准备工作。

② 在实验过程中要严守操作规程，尤其是要严格遵从注意事项，切忌盲目操作，以避免损坏装置或导致实验失败。

③ 严格控制实验条件，仔细观察实验现象，并在编有页码和日期的实验记录本上详细记录原始数据。

④ 对实验中遇到的问题要独立思考，设法解决。若实在困难则应请指导教师协助解决。

⑤ 实验完毕后，将实验数据交指导教师检查，并清理实验桌，洗净并核对仪器，经指导教师同意后才能离开实验室。

（3）实验报告　实验报告是实验工作的书面总结和评定实验工作的依据。编写实验报告是学生分析、归纳实验数据，讨论实验结果，并把在实验中获得的感性认识上升为理性认识的过程。因此，写好实验报告，可以锻炼和培养学生分析问题与解决问题的能力。在完成实验报告时要认真思考，深入研究，做到计算正确、字迹清楚、条理分明。数据处理要求每个学生独立完成。报告要如实反映实验结果，不能拼凑或伪造数据。实验报告可参考以下格式编写：

① 实验目的及简要原理介绍（约300字）。

② 实验　a. 主要仪器药品、实验条件、实验装置原理图等；b. 实验方法的简述及实验的重要步骤等。

③ 实验数据记录及处理　a. 列表已整理的实验数据及进一步运算后的结果；b. 按各项实验的要求进行绘图及计算处理（可用计算机处理）。

④ 结果分析与讨论　a. 将实验结果与文献值比较，从仪器精度、实验技术及误差（包括传递）分析等所得出的结论，对其实验结果作自我评价；b. 对观察到的特殊现象、实验成败的关键、误差产生的原因等问题进行分析讨论，除完成实验项目后面的思考题外，还应结合自己的实验心得体会分析讨论，也可以对实验提出进一步研究与改进的建议。

实验报告要求用统一的实验报告纸撰写，插图应粘贴在报告纸上，报告要求整洁，文字通顺、清晰；下次实验时交上次的实验报告。

1.2　物理化学实验中的安全防护

物理化学实验的安全防护，是一个关系到培养良好的实验素质、保证实验顺利进行、确保实验者和国家财产安全的重要问题。物理化学实验室里经常遇到高温、低温的实验条件，使用高气压（各种高压气瓶）、低气压（各种真空系统）、高电压、高频的仪器，而且许多精密的自动化设备日益普遍使用，因此需要实验者具备必要的安全防护知识，懂得应采取的预防措施，以及一旦事故发生后应及时采取的处理方法。

化学实验室中有各种实验所必需的试剂与仪器，所以常常潜藏着诸如着火、爆炸、中毒、灼伤、触电等安全隐患，如何来防止这些事故的发生以及发生事故以后又如何处置，这些都是每一个化学实验工作者必须具备的素质。本节主要结合物理化学实验的特点做如下介绍。

1.2.1 安全用电

违章用电常常可能造成人身伤亡、火灾、仪器损坏等严重事故。物理化学实验室使用电器较多，要特别注意用电安全。主要应注意以下几点：

① 使用仪器前要根据仪器标牌上所提供的技术数据正确选用电源（如交流、直流、220V、高压电源、低压电源等），接线要正确牢固；
② 操作仪器时，手要保持干燥，切忌用手触摸电源；
③ 要严格按照说明书使用仪器仪表，没有特殊情况，应避免在使用过程中断电；
④ 安装和拆除接线的操作一定要在断电状态下进行，以防触电和电器短路；
⑤ 实验结束后，应关闭仪器电源，并且关闭仪器接线插座上的电源开关；
⑥ 如遇电线走火，切勿用水或导电的酸碱泡沫灭火器灭火，应立即切断电源，用沙或二氧化碳灭火器灭火。

1.2.2 安全使用化学试剂

化学药品使用安全主要有防毒、防爆、防火、防灼伤四个方面。

(1) **防毒**　化学试剂大多数存在不同程度的毒性，其毒性可以通过呼吸道、消化道、皮肤等进入人体。防毒的关键是尽量减少或杜绝直接接触化学试剂。

实验前应了解所用药品的毒性、性能和相关的防毒保护措施。有毒气体应在通风橱内操作。不要在实验室内吃食物，饮具、餐具不能带入实验室．离开实验室时要洗手。

(2) **防爆**　可燃性气体在实验室中达到爆炸极限浓度时，就可能引起爆炸，因而实验室内要尽量减少可燃性气体的挥发。同时要保持实验室良好的通风。

当实验室内有可燃性气体时．应禁止使用明火，防止电火花产生。有些固体试剂如高价态氧化物、过氧化物等受热或撞击时容易引起爆炸，使用时应按要求进行操作。实验室使用高压容器如氧气、氮气、氢气、二氧化碳钢瓶，一定要在教师指导下使用。

(3) **防火**　实验室防火主要有两方面：第一，防止电器设备或带电系统着火，所以用电一定要按规定操作；第二，防止化学试剂着火。许多有机试剂属易燃品，使用这些试剂时要远离火源。实验室一旦发生火灾，应首先切断电源，使用灭火器或沙子灭火。千万不要用水浇。

(4) **防灼伤**　强酸、强碱、强氧化剂等都会灼伤或腐蚀皮肤，尤其要防止进入眼睛，使用时除了要有适当的防护措施外，学生一定要按规定操作。实验室还有高温灼伤（如电炉、烘箱）和低温冻伤（如干冰、液氮）等，使用时也同样要按照规定操作。

1.2.3 安全使用受压容器

物理化学实验室中受压容器主要指高压储气瓶、真空系统、供气流稳压用的玻璃容器，以及盛放液氮的保温瓶等。

(1) **高压储气瓶的安全防护**　高压储气瓶是由无缝碳素钢或合金钢制成，按其所存储的气体及工作压力分类，如表1.1所示。

表 1.1　标准储气瓶型号分类

气瓶型号	用　　途	工作压力 /kgf·cm^{-2}	试验压力/kgf·cm^{-2}	
			水压试验	气压试验
150	氢、氧、氨、氩、氮、甲烷、压缩空气	150	225	150
125	二氧化碳及纯净水煤气等	125	190	125
30	氨、氯、光气等	30	60	30
66	二氧化硫	6	12	6

注：1kgf·cm^{-2}=98.0665kPa。

《气瓶颜色标志》(GB 7144—1999) 规定了各类气瓶的颜色标志（参见附表12），每个气瓶必须在其肩部刻上制造厂和检验单位的钢印标记。

为了安全使用，各类气瓶应定期送检验单位进行技术检查，一般气瓶至少三年检验一次，充装腐蚀性气体的储气瓶至少每两年检验一次。不合格者应降级使用或予以报废。

(2) 使用储气瓶的注意事项

① 气瓶放置要求　气瓶应存放在阴凉、干燥、远离热源（如夏日应避免日晒，冬天与暖气片隔开，平时不要靠近炉火等）的地方，并将气瓶固定在稳固的支架、实验桌或墙壁上，防止受外来撞击和意外跌倒。易燃气体气瓶（如氢气瓶等）的放置房间，原则上不应有明火或电火花产生，确实难以做到时应该采取必要的防护措施。

② 使用时安装减压器（阀）　气瓶使用时要通过减压器使气体压力降至实验所需范围。安装减压器前应确定其连接尺寸规格是否与气瓶接头相符，接头处需用专用垫圈。一般可燃性气体气瓶接头的螺纹是反向的左牙纹，不燃性或助燃性气体气瓶接头的螺纹是正向的右牙纹。有些气瓶需使用专用减压器（如氨气瓶），各种减压器一般不得混用。减压器都装有安全阀，它是保护减压器安全使用的装置，也是减压器出现故障的信号装置。减压器的安全阀应调节到能接受气体的系统或容器的最大工作压力。

③ 气瓶操作要点　气瓶需要搬运或移动时，应拆除减压器，旋上瓶帽，并使用专门的搬移车。启开或关闭气瓶时，实验者应站在减压阀接管的侧面，不许将头和身体对准阀门出口。气瓶启开使用时，应首先检查接头连接处和管道是否漏气，确认无误后方可继续使用。使用可燃性气瓶时，更要防止漏气或将用过的气体排放在室内，并保持实验室通风良好。使用氧气瓶时，严禁气瓶接触油脂，实验者的手、衣服和工具上也不得沾有油脂，因为高压氧气与油脂相遇会引起燃烧。氧气瓶使用时如发现漏气，不得用麻、棉等物去堵漏，以防发生燃烧事故。使用氢气瓶时，导管处应加防止回火装置。气瓶内气体不应全部用尽，应留有不少于 1kgf·cm^{-2} 的压力气体，并在气瓶上标有已用完的记号。

1.2.4　受压玻璃仪器的安全防护

物理化学实验室的受压玻璃仪器包括供高压或真空试验用的玻璃仪器，装载水银的容器、压力计，以及各种保温容器等。使用这类仪器时必须注意：

① 受压玻璃仪器的器壁应足够坚固，不能用薄壁材料或平底烧瓶之类的器皿。

② 供气流稳压用的玻璃稳压瓶，其外壳应裹有布套或细网套。

③ 物理化学实验中常用液氮作为获得低温的手段，在将液氮注入真空容器时，要注意真空容器可能发生破裂，不要把脸靠近容器的正上方。

④ 装载水银的 U 形压力计或容器，要防止系统压力变动过于剧烈、使用时玻璃容器破裂而造成水银散溅到桌上或地上。因此，装载水银的玻璃容器下部应放置搪瓷盘或适当的容器。

⑤ 使用真空玻璃系统时，要注意任何一个活塞的开、闭均会影响系统的其他部分，因此操作时应特别小心，防止在系统内形成高温爆鸣气混合物或让爆鸣气混合物进入高温区。在开启或关闭活塞时，应两手操作，一手握活塞套，另一只手缓缓旋转内塞，勿使玻璃系统各部分产生力矩，以免扭裂。在用真空系统进行低温吸附实验时，当吸附剂吸附大量吸附质气体后，不能先将装有液氮的保温瓶从盛放吸附剂的样品管处移去，而应先启动机械泵对系统进行抽空，然后移去保温瓶。因为一旦先移去低温的保温瓶，又不及时对系统抽空，则被吸附的吸附质气体，由于吸附剂温度的升高，会大量脱附出来，导致系统压力过大，使 U 形压力计中的水银冲出或引起封闭玻璃系统爆裂。

1.2.5　防止环境污染

由于化学试剂大多具有一定的毒性，随意排放会造成环境污染。实验后，废弃药品尽量回收，不能回收的按要求进行处理，符合环保要求后才能排放。

由于汞在物理化学实验室中的应用很普遍，如气压计、水银温度计、U 形汞压差计以及含汞电极等都要用到汞。汞蒸气的最大安全浓度为 0.01×10^{-6} kg·m^{-3}，然而汞在常温下可挥发出的蒸气浓度是安全浓度的一百多倍。汞蒸气可通过呼吸或皮肤直接吸收而使人体中毒，所以防止汞污染尤为重要。

使用汞时，应注意不要将汞直接暴露于空气中，在 U 形汞压差计等汞面上应加水或其他液体，尽量避免汞蒸气外逸。盛汞的容器应有足够的机械强度，以免容器破裂。在实验中要尽量避免因水银温度计、U 形汞压差计等以及含汞电极的人为损坏而造成汞污染。若有汞掉落在桌上或地面上时，应用吸汞管尽可能地将汞珠收集起来，然后用硫黄覆盖在汞掉落的地方，摩擦使之生成 HgS 并清除。

1.3　测量误差及分析

物理化学实验通常是在一定条件下测定系统的一种或几种物理量的大小，然后用计算或作图的方法求得所需的实验结果。在测定过程中，即使采用最可靠的测量方法、使用最精密的仪器、由技术很熟练的人员进行操作，也不可能得到绝对准确的结果。因为在任何测量过程中，误差是客观存在的。因此我们应该了解实验过程中误差产生的原因及出现的规律，以便采取相应措施减少误差。另外，需要对测试数据进行正确的处理，以获得最可靠的数据信息。在本实验课中除了学习误差的基本概念外，还要求学生能计算间接测量的误差，掌握作图方法，以及正确表达测量结果等。这些内容是物理化学实验技能的必备素质，一定要给予足够的重视。

1.3.1　基本概念

1.3.1.1　误差与准确度

（1）误差　误差是指测定值 x_i 与真值 a 之差。误差的大小可用绝对误差 E 和相对误差 E_r 表示，即

$$E = x_i - a \tag{1.1}$$

$$E_r = \frac{x_i - a}{a} \times 100\% \tag{1.2}$$

相对误差表示误差占真值的百分率。

例如，分析天平称量物体的质量分别为 1.6380g 和 0.1637g，假如两者的真实质量分别为 1.6381g 和 0.1638g，则两者称量的绝对误差分别为

$$E = 1.6380 - 1.6381 = -0.0001(\text{g})$$

$$E = 0.1637 - 0.1638 = -0.0001(\text{g})$$

两者称量的相对误差分别为

$$E_r = \frac{-0.0001}{1.6381} \times 100\% = -0.006\%$$

$$E_r = \frac{-0.0001}{0.1638} \times 100\% = -0.06\%$$

从上例可知，绝对误差相等，相对误差并不一定相同。第一个称量结果的相对误差为第二个称量结果的 1/10。由此我们得出这样的结论：同样的绝对误差，当被测定的量较大时，相对误差就比较小，测定的准确度也就比较高。因此，用相对误差来表示各种情况下测定结果的准确度更为确切些。

绝对误差和相对误差都有正值和负值。正值表示测量结果偏高，负值表示测量结果偏低。实际测量中，真值实际上是无法获得的，人们常常用纯物质理论值、国家标准局提供的标准参考物质的证书上给出的数值或多次测定结果的平均值当作真值。

（2）准确度（正确度） 它反映了由系统误差引起的测量值的偏离程度。系统误差愈小，测量结果的准确度愈高。准确度的定义为：

$$b = \frac{1}{n} \sum_{i=1}^{n} |x_i - a| \tag{1.3}$$

式中，n 为测量次数；x_i 为第 i 次的测量值；a 为真值。

由于在大多数物理化学实验中，真值 a 是我们要求的测定的结果，因此 a 值就很难得到。但一般可近似地用标准值 $x_{标}$ 来代替 a（$x_{标}$ 是用其他更可靠的方法测出的值，也可用文献手册的公认值代替）。此时测量的准确度可近似表示为

$$b = \frac{1}{n} \sum_{i=1}^{n} |x_i - x_{标}| \tag{1.4}$$

如果实验结果没有 $x_{标}$，则可用不同的实验方法经过多次测量结果的平均值代替 $x_{标}$。但最终结果还需实践的检验。

1.3.1.2 偏差与精密度

偏差是指个别测定结果与几次测定结果的平均值 $\overline{x_i}$ 之间的差别。与误差相似，偏差也有绝对偏差 d_i 和相对偏差 d_r 之分。测定结果与平均值之差为绝对偏差，绝对偏差在平均值中所占的百分率或千分率为相对偏差，即

$$d_i = x_i - \overline{x} \tag{1.5}$$

$$d_r = \frac{|x_i - \overline{x}|}{\overline{x}} \times 100\% \tag{1.6}$$

各偏差值的绝对值的平均值，称为单次测定的平均偏差 $\overline{d}$，又称为算术平均偏差，即

$$\overline{d} = \frac{1}{n} \sum_{i=1}^{n} |d_i| = \frac{1}{n} \sum_{i=1}^{n} |x_i - \overline{x}| \tag{1.7}$$

单次测定的相对平均偏差 $\overline{d_r}$ 表示为

$$\overline{d_r} = \frac{\overline{d}}{\overline{x}} \times 100\% \tag{1.8}$$

精密度反映了同一物理量多次测量结果的彼此符合程度。反映了偶然误差对测量结果的影响。偶然误差愈小，测量值彼此愈符合，则精密度愈高。精密度的大小还反映了测量结果的有效数字位数多少（与所有测量仪器的分辨能力有关）。如果测量结果的重复性高且有效数字位数多，则可以认为精密度高。精密度的大小常用偏差来表示。

1.3.1.3　精确度及准确度与精密度的关系

精确度反映的是由系统误差和偶然误差共同引起的测量值对真值的偏离程度，即测量结果与其真值符合程度的量度。它与误差的大小相对应。误差大，精确度低；误差小，精确度高。由于任何实验测量值都无法消除全部误差，故一般的情况下实验测量得到真值是不可能的，故常用多次测量结果的算术平均值或用文献手册所查的公认值代替真值。精确度包括准确度和精密度两部分含义。精确度高表示准确度和精密度都高。

用一个例子来说明精确度、准确度和精密度的关系。甲、乙、丙三人测量某一物理量，其结果如图1.1所示，图中所示的测量结果就表示了精密度与准确度和精确度的关系。

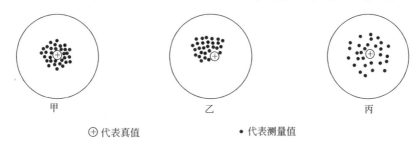

图1.1　测量精密度与准确度和精确度

甲：系统误差和偶然误差都小，精密度、准确度都高即精确度高；乙：系统误差大，
而偶然误差小，即精密度高，准确度低即精确度低；丙：系统误差小，
而偶然误差大，即准确度高，精密度低即精确度低

1.3.1.4　误差产生的原因及分类

一般测量误差可分为系统误差、偶然误差和过失误差。

（1）系统误差　系统误差是由于某种特殊原因引起的误差。它对测量结果的影响是固定的或是有规律变化的。它使测量结果总是偏向一方，即总是偏大或偏小。测量次数的增加并不能使之消除。

系统误差按产生原因的不同可分类如下。

① 仪器误差　仪器误差是指在进行测量时所使用的测量设备或仪器本身固有的各种因素的影响产生的误差。测量装置的技术指标，如准确度、灵敏度、最小分度值、变差值及稳定度等的好坏取决于测量装置的结构、设计、所用元器件的性能、零部件材料的性能，加工制造和装配的技术水平等因素。在设计和制造各种测量仪器时，只能根据现有的条件与可能提出的实际要求，尽量减少误差，而与理想的要求总会有一定的差距，所以在测量过程中使用装置、设备和仪器仪表时，无论怎样满足规定的使用条件，无论怎样细心操作，总会使测量值产生误差。

② 试剂误差　这是在化学实验中，所用试剂纯度不够而引起的误差。在某些情况下，试剂所含杂质可能给实验结果带来严重的影响。

③ 方法误差　它是由于所采用的测量原理或测量方法本身所产生的测量误差。构成此

类误差的来源，常遇到的有：

　　a. 对被测对象的有关知识研究得不够充分，不能全面地考虑一些因素对测量所造成的影响；

　　b. 受客观条件及技术水平的限制；

　　c. 应用的测量原理本身就是近似性的或忽略了一些在测量过程中实际在起作用的因素；

　　d. 用接触测量破坏了被测对象的原始状况；

　　e. 用静态的测量方法解决动态对象测量。

　　只有用多种方法测得的同一数据相一致时，才可认为方法误差已基本消除、结果是可取的。如元素原子量总是用多种方法测定而确定的。

　　④ 个人误差　个人误差是由进行测量的操作人员的习惯和特点引起的误差。主要是因为测量人员感觉器官的分辨能力、反应滞后、习惯感觉等因素而引起的观测误差。如记录某一信号的时间总是提前或滞后，读取仪表时眼睛位置总是偏向一边，判定滴定终点的颜色各人不同等。

　　⑤ 环境误差　因为周围环境因素对测量的影响，而使测量产生的误差。这些影响因素存在于测量系统之外，但对测量系统会直接或间接发生作用，例如温度、湿度、大气压、电场、磁场、机械振动、加速度、地心引力、声响、光照、灰尘、各种射线或电磁波等。这些因素在不同的测量过程中，对测量产生的影响程度可能不同。它们不但能影响测量系统产生测量误差，有时也能引起被测量系统的变化，严重时甚至会造成测量设备的毁坏或使测量难以进行。为了区分环境误差和仪器误差，人为地确定所谓标准环境（基准条件），或在产品铭牌及使用说明书上规定测量仪器的使用条件。在基准条件下进行测量所产生的测量误差基本上认为是测量仪器的固有误差（仪器误差）。若使测量仪器在超出基准条件规定的环境下工作，因为环境因素的影响，造成测量误差的增大，这种测量误差的增加量，称为仪器的附加误差，也就是环境误差。因此仪器在满足规定的条件下进行测量，所获测量值的误差不应超过铭牌或说明书中给出的误差值。有些仪表还给出随环境条件变化而改变的环境误差值。

　　上述五种测量误差的来源是从参加测量的四个环节，即人员、设备、方法和条件概括出来的。在具体测量过程中，各因素对测量的影响程度有所不同，甚至达到某一因素造成的测量误差可以忽略的程度，但测量得到的测得值总会带有测量误差是不容怀疑的。

　　系统误差影响了测量结果的准确程度。系统误差的数值可能比较大。必须消除系统误差的影响，才能有效地提高测量的精确度。实验工作者的重要任务之一就是发现系统误差的存在，找出系统误差的主要来源，选择有效的消除或减少系统误差的办法。通常可采用几种不同的实验技术或采用不同的实验方法。或改变实验条件，调换仪器，提高化学试剂的纯度等以确定有无系统误差的存在，并设法消除或使之减少。因此，单凭一种方法所得结果往往不是十分可靠的，只有由不同实验者、用不同的方法、不同的仪器得到相符的数据，才能认为系统误差基本消除。

　　(2) 偶然误差　在实验时即使采用了最先进的仪器、选择了最恰当的方法，经过了十分精细的观测消除了系统误差，在同一条件下对一个物理量进行重复测量时，所测得的数据也不可能每次相同，在数据的末一位或末二位数字上仍会有差异，即存在着一定的误差，这种误差称为偶然误差。偶然误差是由测量过程中一系列偶然因素（实验者不能严格控制的因素，如外界条件、实验者心理状态、仪器结构不稳定等）引起的。偶然误差在测量时不可能消除或估计出来，但是它服从统计规律。实践经验和概率论都证明了，在相同条件下，多次测量同一个物理量，当测量次数足够多时，出现偶然误差数值相等、符号相反的测量结果的概率近乎相等。通过增加测量次数可使偶然误差减小到某种需要的程度。偶然误差决定测量结果的精密度。

偶然误差的出现在表面上看没有确定的规律，即前一误差出现后，不能料想下一个测量误差的大小和方向，但就其总体而言，具有统计规律性。

（3）过失误差　过失误差是由于实验者的过失或错误引起的误差，如读移液管刻度出现错误、计算错误、记录写错等。含有过失误差的测量值是坏值，应该从结果中将它剔除。过失误差无规律可循，只要工作仔细、加强责任心就可以避免。防止过失误差还可以使用校核法，即用别的方法或仪器对测量值进行近似测量，以判断正式测量的数据是否合理。过失误差在测量中应尽力避免。

系统误差与偶然误差之间虽有着本质的不同。但在一定条件下它们可以互相转化。实际上，我们常把某些具有复杂规律的系统误差看作偶然误差，采用统计的方法来处理。不少系统误差的出现均带有随机性。例如，在用天平称量时，每个砝码都存在着大小不等、符号不同的系统误差。这种系统误差的综合效果，对每次称量是不相同的，它具有很大的偶然性。因此，在这种情况下，我们也可把这种系统误差作为偶然误差来处理。

对按准确度划分等级的仪器来说，同一级别的仪器中每个仪器具有的系统误差是随机的，或大或小、或正或负，彼此都不一样。如一批容量瓶中，每个容量瓶的系统误差不一定相同，它们之间的差别是随机的，这种误差属于偶然误差。当使用其中某一个容量瓶时，这种随机的偶然误差又转化为系统误差。我们可通过校核，确定其系统误差的大小。如不校核或未被发现，仍然当作偶然误差来处理也是常有之事。有时，系统误差与偶然误差的区分也取决于时间因素。在短期内是基本不变的系统误差，但时间一长，则可能出现随机变化的偶然误差。

1.3.2　误差的表示方法

（1）算术平均值 $\bar{x}$

$$\bar{x} = \frac{x_1 + x_2 + x_3 + \cdots + x_n}{n} \tag{1.9}$$

式中，x_1，x_2，x_3，$\cdots$，x_n 为测量值；n 为测量次数。

（2）绝对误差 E 和绝对偏差 d

$$E_i = x_i - a \tag{1.10}$$

$$d_i = x_i - \bar{x} \tag{1.11}$$

式中，a 为真实值。

（3）平均误差 δ

$$\delta = \frac{\sum |d_i|}{n} \tag{1.12}$$

式中，$i = 1, 2, 3, \cdots, n$；$d_1 = x_1 - \bar{x}$，$d_2 = x_2 - \bar{x}$，$d_3 = x_3 - \bar{x}$，$\cdots$，$d_n = x_n - \bar{x}$。

（4）标准误差 σ 和标准偏差 s　标准误差 σ 也称均方根误差，其定义为：

$$\sigma = \sqrt{\frac{\sum_{i=1}^{n} E_i^2}{n-1}} \quad (n > 30) \tag{1.13}$$

$$s = \sqrt{\frac{\sum_{i=1}^{n} d_i^2}{n-1}} \quad (n < 30) \tag{1.14}$$

计算标准误差 σ 和标准偏差 s 是评定精密度的最好方法，在现代科学中广为采用。测量结果表示为 $x\pm\sigma$ 或 $x\pm s$。

(5) 或然误差 P　或然误差 P 的意义是：在一组测量中若不计正负号，误差大于 P 的测量值与误差小于 P 的测量值，将各占测量次数的 50%，即误差落在 $+P$ 与 $-P$ 之间的测量次数，占总测量数的一半。

以上三种误差之间的关系为：$P:\delta:\sigma=0.675:0.794:1.00$。

(6) 相对误差

$$\text{相对误差}=\frac{\Delta x}{a}\times 100\% (\Delta x \text{ 可为 } P, \delta, \sigma) \tag{1.15}$$

$$\text{相对偏差}=\frac{\Delta x}{\bar{x}}\times 100\% (\Delta x \text{ 可为 } P, \delta, \sigma) \tag{1.16}$$

(7) 极差 R　极差 R 是指一组测定数据中，最大值和最小值之差。可用它表示误差范围。极差又称为范围误差，即

$$R=\max(x_1, x_2, \cdots, x_n)-\min(x_1, x_2, \cdots, x_n), \tag{1.17}$$

式中，$\max(x_1, x_2, \cdots, x_n)$ 和 $\min(x_1, x_2, \cdots, x_n)$ 分别表示 $x_1, x_2, \cdots, x_n$ 中最大和最小的数值。

绝对误差的单位与测量值的单位相同，相对误差是无量纲量。对于同一量的测量，绝对误差可以确定其测量精度的高低。而对于不同量的测量，只能采用相对误差来评定才较为确切。平均误差的优点是计算简便，但不能肯定 x_i 相对 $\bar{x}$ 是偏高还是偏低，用这种误差表示时，可能会把质量不高的测量值掩盖住。相对误差可用于比较各种测量的精度，评价测量结果的质量。标准误差对一组测量中的较大误差或较小误差感觉比较灵敏，其测量结果的精度常用 $(\bar{x}\pm\sigma)$ 或 $(\bar{x}\pm\delta)$ 来表示，σ 值或 δ 值越小，表示测量的精密度越好。因此近代科学中多采用标准误差来表示测量的精度。极差虽然能反映测定实际数据的波动范围，但没有充分利用数据提供的情报，不能全面、科学地反映测定数据的质量。但由于它计算简便，在快速检验中得到广泛应用。

1.3.3 偶然误差的统计规律及其应用

在消除了系统误差和杜绝了过失误差后，测量的误差只有偶然误差。误差理论主要研究偶然误差的特性及其应用。

1.3.3.1 偶然误差的基本特性

① 同样大小的正误差和负误差的出现次数相等，当测量次数足够多时，$\bar{x}\to a$。

② 测量结果中误差小的值出现次数多（概率大），而误差大的值出现次数少（概率小）。

③ 绝对值很大的误差不会出现，即偶然误差有一定的界限。

1.3.3.2 偶然误差的规律

根据偶然误差的基本特性，如横轴表示测量值 x，纵轴表示各个偶然误差出现的频率，则得图 1.2 的偶然误差分布曲线，即正态分布曲线。偶然误差的分布曲线反映了误差的大小与其出现的频率的关系。

横轴表示偶然误差时，曲线最高点对应的横坐

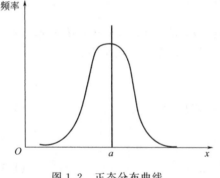

图 1.2　正态分布曲线

标表示误差为零，横轴表示测量值时，曲线最高点对应的横坐标表示真值 a。

1.3.3.3 高斯误差方程

偶然误差的分布曲线反映了误差的大小与其出现概率的关系，1795 年高斯确定该函数的形式为

$$\varphi(x)=\frac{1}{\sigma\sqrt{2\pi}}\exp\left[-\frac{(x-a)^2}{2\sigma^2}\right] \tag{1.18}$$

$$\varphi(x)=\frac{h}{\sqrt{\pi}}\exp\left[-h^2(x-a)\right] \tag{1.19}$$

式中，$\varphi(x)$ 为正态概率密度函数；x 为测量值；a 为真值；σ 为总体标准偏差；h 为精密度指数。

当 $x-a=0$ 时，$\varphi(x)$ 值最大，可表示为

$$\varphi(a)=\frac{1}{\sigma\sqrt{2\pi}}=\frac{h}{\sqrt{\pi}} \tag{1.20}$$

从上式可知，$\varphi(x)$ 与 σ 成反比，与 h 成正比。σ 值越小（h 值越大），误差分布曲线越尖，较小误差出现的概率越大，测量的精密度也越高。反之，σ 值越大（h 值越小），误差分布曲线越平坦，较大的误差出现的概率越大。若以横坐标表示偶然误差 σ 或测量值 x，纵轴表示 $\varphi(x)$，作图则得图 1.3 的不同精密度的误差分布曲线。

横轴表示偶然误差时，曲线最高点对应的横坐标表示误差为零；横轴表示测量值时，曲线最高点对应的横坐标表示真值 a。图 1.3 中三条误差分布曲线的精密度不同，标准误差也不同，$\sigma_1<\sigma_2<\sigma_3$。

1.3.3.4 偶然误差的概率

高斯误差方程中有两个变量：误差 $(x-a)$ 和 σ，为了用一个变量误差的函数形式，令

$$u=\frac{x-a}{\sigma} \tag{1.21}$$

则有

$$\varphi(u)=\frac{1}{\sqrt{2\pi}}e^{-\frac{u^2}{2}} \tag{1.22}$$

如以偶然误差出现的次数为纵轴，u 为横轴，作图可得标准正态分布曲线。它可将不同精密度测量的正态分布曲线统一为一条曲线，但各自的 u 值大小不同，如图 1.4 所示。

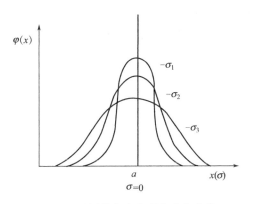

图 1.3 不同精密度的误差分布曲线

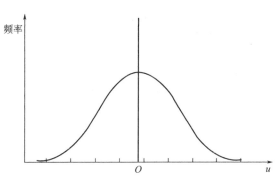

图 1.4 正态分布曲线

根据 $\varphi(u)$ 可计算某个误差在某一误差范围出现的概率。在方程式中，u 在 $-\infty$ 和 $+\infty$ 之间的积分为 1，即

$$\frac{1}{\sqrt{2\pi}}\int_{-\infty}^{+\infty} e^{-\frac{u^2}{2}} du = 1 \tag{1.23}$$

在实际应用时用下式求误差的概率 P：

$$P = \frac{1}{\sqrt{2\pi}}\int_{-\infty}^{+\infty} e^{-\frac{u^2}{2}} du \tag{1.24}$$

对于标准正态分布，样本值大于边界值 K_a 时的概率 $P_u \geq K_a$：

$$P_u \geq K_a = \frac{1}{\sqrt{2\pi}}\int_{K_a}^{+\infty} e^{-\frac{1}{2}u^2} du = \alpha \tag{1.25}$$

为了应用方便，将上式制成表来计算不同 u 值误差的概率。下面选列几个常用数据的正态分布误差的概率。从表 1.2 中可知：

表 1.2　几个常用正态分布误差的概率

误差以 σ 为单位	概率<$|u|$的正负误差	概率>$|u|$的正负误差
0.000	0.000	1.000
0.675	0.500	0.500
1.000	0.683	0.317
1.645	0.900	0.100
1.960	0.950	0.050
2.000	0.955	0.045
2.576	0.990	0.010
3.000	0.997	0.003
3.219	0.999	0.001

误差在 $\pm 0.675\sigma$ 范围的概率是 50%；

误差在 $\pm\sigma$ 范围的概率是 68.3%；

误差在 $\pm 2\sigma$ 范围的概率是 95.5%；

误差在 $\pm 3\sigma$ 范围的概率是 99.7%。

在相同条件下，测量次数增加则平均误差减少。测量四次比测量一次的准确度高出三倍，测量九次比测量一次的准确度也高出三倍。一般四次测量的平均值就能基本满足要求。

对有限次测量平均值的取舍，可采用 $\pm 3\sigma$ 或 $\pm 3s$ 规则，即测量的平均值在 $\bar{x} \pm 3\sigma$ 或 $\bar{x} \pm 3s$ 范围内的值可认为有 99% 的可靠性。

1.3.3.5　小量样本统计学 t 分布及其应用

(1) t 分布　前面讨论的误差正态分布理论是从大量数据中推论出来的，在实际测量中，测量次数不可能无限多。在等精密度的多次测量中，测量次数大于 30 个，称为大样本测量。可用 $\bar{x}$ 代表最佳值，用标准偏差 s 代替标准误差 σ。但在物理化学实验中，一般对物理量往往只进行少数几次测量，称为小样本测量。由于测量次数少，偶然误差正态分布理论不能直接用于小样本测量的检验。对于小样本测量，样本值服从 t 分布，用下式表示：

$$t = \frac{\bar{x} - a}{s_{\bar{x}}} = \frac{\bar{x} - a}{s/\sqrt{n}} = (\bar{x} - a)\frac{\sqrt{n}}{s} \tag{1.26}$$

t 分布的概率密度由 t 分布密度函数 $\varphi(t)$ 给出：

$$\varphi(t) = \frac{1}{\sqrt{\pi f}} \times \frac{\Gamma\left(\frac{f+1}{2}\right)}{\Gamma(f/2)}\left(1 + \frac{t^2}{f}\right)^{-\frac{f+1}{2}} \tag{1.27}$$

t 值用来量度小样本测量误差，误差分布类似正态分布，称之为"t分布"，如图 1.5 所示，图中 f 为自由度，$f=n-1$。由图可见：测量次数越少（f 越小），曲线越扁平；当 f 无限大时，t 分布与正态分布曲线相同，此时 $t=u$；当 $f>20$ 时，t 分布曲线和正态分布曲线相似；当 $f<10$ 时，t 分布和正态分布曲线相差较大。

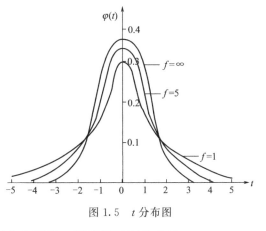

图 1.5　t 分布图

应用 t 分布时，把 t 分布列成表。从表中查出不同置信水平下和不同自由度的临界 t 值（t 分布表本书未列，可参考相关教材）。

（2）t 分布的应用　判断一个实验方法是否准确可靠，可采用以下的方法：
① 计算用该实验方法求得结果的平均值 $\bar{x}$；
② 用式(1.14)求 s 值；
③ 用式(1.26)求 t 值；
④ 查 t 分布表：当 $t_表 < t_算$ 时，该方法不够准确，有系统误差；当 $t_表 > t_算$ 时，该方法是可靠的，结果也可靠。

1.3.3.6　可疑值的取舍

在一组测量值中，如发现其中某个测量值明显比其他的测量值大或小。对于这个测量值，既不能轻易保留，也不能随意舍弃，可用如下方法处理。

（1）3σ 准则　即测量的平均值在 $x \pm 3\sigma$ 或 $x \pm 3s$ 范围内的值可认为有 99% 的可靠性。

（2）Q 检验法
① 将测量值从小到大按顺序排列出来，如 $x_1 < x_2 < x_3 \cdots < x_n$，其中有两个值怀疑其准确性；
② 计算 Q 值，以测量值的最大值与最小值之差为分母，可疑值与其相邻值之差为分子计算；
③ 将计算的 Q 值与 Q 表值比较，若 $Q \geqslant Q_表$，则其值应舍弃；若 $Q < Q_表$，则其值应保留。表 1.3 为 Q 值表。

表 1.3　90%，95% 的 Q 值

Q 值＼n	3	4	5	6	7	8	9	10
90%	0.94	0.76	0.64	0.56	0.51	0.47	0.44	0.41
95%	1.53	1.05	0.86	0.76	0.69	0.64	0.60	0.58

（3）t 检验法　t 检验法是应用 t 分布测定的平均值和标准值相比较，或不同实验者，不同实验方法测定的平均值之间的比较。

1.3.4　间接测量结果的误差计算

在大多数情况下，要对几个物理量进行测量，通过函数关系进行计算，才能得到所需的结果，这就是间接测量。在间接测量中，每个直接测量值的精确度都会影响最后结果的精确度。因而间接测量结果的误差要通过直接测量值的误差进行计算得到。

1.3.4.1　间接测量结果的平均误差

设直接测量的数据为 x 及 y，其平均误差为 Δx 及 Δy，而最后结果为 z，其函数关系可

表示为
$$Z = F(x, y) \tag{1.28}$$

微分求 Δz 得
$$\Delta z = \left(\frac{\partial F}{\partial x}\right)_y \Delta x + \left(\frac{\partial F}{\partial y}\right)_x \Delta y \tag{1.29}$$

式(1.29)是计算间接测量结果的平均误差的基本公式。有关其他误差的计算可参考表1.4进行。

表 1.4　部分函数的其他误差的计算公式

函数关系	绝对误差	相对误差
$z = x + y$	$\pm(|\Delta x| + |\Delta y|)$	$\pm\dfrac{|\Delta x| + |\Delta y|}{x+y}$
$z = x - y$	$\pm(|\Delta x| + |\Delta y|)$	$\pm\dfrac{|\Delta x| + |\Delta y|}{x-y}$
$z = xy$	$\pm(x|\Delta x| + y|\Delta y|)$	$\pm\left(\dfrac{|\Delta x|}{x} + \dfrac{|\Delta y|}{y}\right)$
$z = \dfrac{x}{y}$	$\pm\dfrac{x|\Delta x| + y|\Delta y|}{y^2}$	$\pm\left(\dfrac{|\Delta x|}{x} + \dfrac{|\Delta y|}{y}\right)$
$z = x^n$	$\pm(nx^{n-1}\Delta x)$	$\pm\left(n\dfrac{\Delta x}{x}\right)$
$z = \ln x$	$\pm\dfrac{\Delta x}{x}$	$\pm\left(n\dfrac{\Delta x}{x\ln x}\right)$

例如：$z = x + y$，其相对误差为
$$\frac{\Delta z}{z} = \frac{\Delta x}{x} + \frac{\Delta y}{y} \tag{1.30}$$

百分误差为
$$\frac{\Delta z}{z} \times 100\% = \frac{\Delta x}{x} \times 100\% + \frac{\Delta y}{y} \times 100\% \tag{1.31}$$

下面将举例加以说明。

【例 1.1】 在进行凝固点降低法测分子量实验时，用下式计算 M：
$$M = \frac{1000 K_f W_B}{W_A \Delta T_1} = \frac{1000 K_f W_B}{W_A (T_f^* - T_f)}$$

这里直接测量的数值为 W_B，W_A，T_f^*，T_f。

溶质质量 $W_B = 0.300\text{g}$，在分析天平上称量的绝对误差 $\Delta W_B = 0.0002\text{g}$，溶剂质量 $W_A = 20.00\text{g}$，在粗天平上称量的绝对误差 $\Delta W_A = 0.05\text{g}$。

测量凝固点用贝克曼温度计，其读数误差为 $0.002℃$，测出溶剂的凝固点 T_f^* 三次，分别为 $5.801℃$，$5.790℃$，$5.802℃$，则
$$\overline{T_f^*} = \frac{5.801 + 5.790 + 5.802}{3} = 5.798(℃)$$

每次测量误差为：
$$\Delta T_{11}^* = |5.798 - 5.801| = 0.003(℃)$$
$$\Delta T_{12}^* = |5.798 - 5.790| = 0.008(℃)$$
$$\Delta T_{13}^* = |5.798 - 5.802| = 0.004(℃)$$

平均误差为：
$$\overline{T_f^*} = \pm\frac{0.003 + 0.008 + 0.004}{3} = \pm 0.005(℃)$$

同样测出溶液的凝固点三次，分别为 5.500℃，5.4925℃，5.504℃。用同样的方法计算得出 $\overline{T_f}$=5.500℃、$\Delta \overline{T_f}$=0.003℃。这样凝固点降低数值为

$$\Delta T_f = T_f^* - T_f = (5.798 \pm 0.005) - (5.500 \pm 0.003) = 0.298 \pm 0.008(℃)$$

其相对误差为：

$$\frac{\Delta(T_f)}{\Delta T_f} = \frac{0.008}{0.298} = 0.027$$

$$\frac{\Delta W_B}{W_B} = \frac{0.0002}{0.3} = 6.67 \times 10^{-4}$$

$$\frac{\Delta W_A}{W_A} = \frac{0.05}{20} = 25 \times 10^{-4}$$

而测定摩尔质量 M 的相对误差将为

$$\frac{\Delta M}{M} = \frac{\Delta W_A}{W_A} + \frac{\Delta W_B}{W_B} + \frac{\Delta(T_f)}{\Delta T_f}$$
$$= \pm(6.67 \times 10^{-4} + 25 \times 10^{-4} + 2.7 \times 10^{-2})$$
$$= \pm 0.030$$

因此，在凝固点降低法测分子量时的最大相对误差为 3.0%。这一计算表明在用凝固点降低法测分子量时，相对误差决定于测量温度的精确度。根据公式可知，若增加溶质质量，ΔT_1 可以增大，相对误差可以减小，但凝固点降低法测分子量的公式只是在稀溶液下才是正确的。在增加溶质质量减小相对误差的同时却增大了系统误差。计算结果表明，提高称量的精确度并不能增加测量摩尔质量精确度。过分地强调称量的精确度（用分析天平称量溶剂的质量）是不适宜的。而影响测量摩尔质量精确度的关键在于温度的测量。可见，事先了解间接测量时各个测量值的大致误差范围及其影响，就能指导我们选择正确的实验方法，选用精密度合适的仪器。抓住影响误差的关键因素，使测量结果的误差在允许的范围内。

1.3.4.2 间接测量结果的标准误差

设直接测量的数据为 x 及 y，其标准误差为 dx 及 dy，而最后结果为 z，其函数关系可表示为

$$Z = F(x \cdot y) \tag{1.32}$$

则函数 z 的标准误差为

$$\sigma = \sqrt{\left(\frac{\partial z}{\partial x}\right)^2 \sigma_x^2 + \left(\frac{\partial z}{\partial y}\right)^2 \sigma_y^2} \tag{1.33}$$

部分常用函数的标准误差公式列于表 1.5。

表 1.5 部分常用函数的标准误差公式

函数关系	绝对误差	相对误差
$z = x \pm y$	$\pm \sqrt{\sigma_x^2 + \sigma_y^2}$	$\pm \frac{1}{\|x \pm y\|}\sqrt{\sigma_x^2 + \sigma_y^2}$
$z = x \cdot y$	$\pm \sqrt{y^2 \sigma_x^2 + x^2 \sigma_y^2}$	$\pm \sqrt{\frac{\sigma_x^2}{x^2} + \frac{\sigma_y^2}{y}}$
$z = \frac{x}{y}$	$\pm \frac{1}{y}\sqrt{\sigma_x^2 + \frac{x^2}{y^2}\sigma_y^2}$	$\pm \sqrt{\frac{\sigma_x^2}{x^2} + \frac{\sigma_y^2}{y}}$
$z = x^n$	$\pm n x^{n-1} \sigma_x$	$\pm \frac{n}{x} \sigma_x$
$z = \ln x$	$\pm \frac{\sigma_x}{x}$	$\pm \frac{\sigma_x}{x \ln x}$

【例 1.2】 溶质的摩尔质量 M 可由溶液的沸点升高值 ΔT_b 测定。设以苯为溶剂，以萘为溶质，用贝克曼温度计测得纯苯的沸点为 (2.975 ± 0.003)℃，而溶液中含苯 $W_A=(87.0\pm0.1)$g，含萘 $W_B=(1.054\pm0.001)$g，溶液沸点为 (3.210 ± 0.003)℃。试用下列公式计算萘的摩尔质量及标准误差：

$$M=2.53\times\frac{1000W_B}{W_A\Delta T_b}$$

由函数的标准误差公式得出

$$\sigma_M=\sqrt{\left(\frac{\partial M}{\partial W_B}\right)^2\sigma_B^2+\left(\frac{\partial M}{\partial W_A}\right)^2\sigma_A^2+\left(\frac{\partial M}{\partial \Delta T_b}\right)^2\sigma^2\Delta T_b}$$

其中：

$$\frac{\partial M}{\partial W_B}=\frac{2.53\times1000}{W_A\Delta T_b}=\frac{2.53\times1000}{87.0\times0.235}=124$$

$$\frac{\partial M}{\partial W_A}=\frac{2.53\times1000W_B}{\Delta T_b}\times\frac{1}{W_A^2}=\frac{2.53\times1000\times1.054}{0.235\times87.0^2}=1.50$$

$$\frac{\partial M}{\partial \Delta T_b}=\frac{2.53\times1000W_B}{W_A}\times\frac{1}{\Delta T_A^2}=\frac{2.53\times1000\times1.054}{87.0\times0.235^2}=555$$

$$\sigma_M=\sqrt{124^2\times0.001^2+1.50^2\times0.1^2+555^2\times(0.003+0.003)^2}=3.3$$

$$M=2.53\times\frac{1000\times1.054}{87.0\times0.235}=130(\text{g}\cdot\text{mol}^{-1})$$

萘的摩尔质量最后表示为：$(130\pm3)\text{g}\cdot\text{mol}^{-1}$。

1.4 实验数据的表达方法

物理化学实验数据的表达及处理方法主要有三种：列表法、图解法和数学方程式法。

1.4.1 列表法

列表法是以列表的方式将实验结果的自变量 x 和因变量 y 的相应数值一一对应列出。该方法的优点是能使全部数据一目了然，便于处理运算、容易检查而减少差错。

列表时应注意以下几点：

① 每一个表都应有简明而完备的名称。

② 表中的每一行或每一列的第一栏要详细地写出数据的名称和量纲；

③ 在每一行（或列）中，数字排列要整齐，位数和小数点要对齐，有效数字的位数要合理；

④ 表中的数据应化成最简单的形式表示，公共的乘方因子应在第一栏的名称下注明；

⑤ 表中的数据应按依次递增或递减排列，缺的数据用"—"表示。

列表法简单易行，不需要特殊图纸（如方格纸）和仪器，形式紧凑，又便于参考比较。在同一表格内，可以同时表示几个变量间的变化情况。实验的原始数据一般采用列表法记录。

表 1.6 是 CO_2 的平衡性质，其形式可作为一般参考。

表 1.6 CO_2 的平衡性质

$t/℃$	T/K	$10^3 K/T$	p/MPa	$\ln(p/MPa)$	$V_m^g/(cm^3 \cdot mol^{-1})$	pV_m^2/RT
−56.60	216.5	4.6179	0.5180	−0.6578	3177.6	0.9142
0.00	273.15	3.6610	3.4853	1.2485	456.97	0.7013
31.04	304.19	3.2874	7.382	1.9990	94.060	0.2745

像 $V_m^g/(cm^3 \cdot mol^{-1})$ 这样的表头，如果嫌它占的地方太宽，可以用相当的其他式子来代替，如可将其写成 $\dfrac{V_m^g}{cm^3 \cdot mol^{-1}}$。

1.4.2 图解法

实验数据图解法是根据解析几何原理，用几何图形将实验数据表示出来。其优点是能直观地表现出实验测得的各数据间的相互关系，并能清楚地显示出所研究问题的变化规律，如极大值、极小值、转折点、周期性、数量变化的速率等，还易于从图上找出所需数据，同时便于数据的分析比较和进一步求得函数关系的数学表达式。因而该方法在物理化学实验的数据表达和处理中应用十分广泛。

1.4.2.1 图解法在物理化学实验中的应用

（1）求内插值　以自变量为横轴，以因变量为纵轴，所得曲线即表示二变量间的定量依赖关系。在曲线所示的范围内，可方便地从曲线上求出任一自变量所对应的因变量的数值。例如，二元液系汽液平衡相图实验中，从不同组成溶液的折射率工作曲线上直接读出某一折射率对应的溶液组成。

（2）求外推值　若测定的物理量不能或不易由实验直接测定，在一定的条件下，将所测量的数据间的函数关系外推至测量范围之外，可获得所需要的数值。例如，过氧化氢分解实验中 H_2O_2 完全分解时量气管的读数 V_∞ 不易直接由实验获得，但可测定不同时刻 t 时量气管的读数 V_t，作 V_t-$1/t$ 图外推至 $1/t$ 等于 0 时的 V_t 即为 V_∞。值得注意的是外推法只有在下列情况下才能应用：

① 在外推的那段范围及其邻近，测量数据间的函数关系是线性关系或可认为是线性关系；

② 外推范围距实际测量范围不能太远；

③ 外推所得的结果与已有的正确经验不能相抵触。

（3）作切线求函数的微商（图解微分法）　从曲线上选定若干点作切线，计算出该点的斜率，即得该点的微商值。例如，利用不同浓度溶液的表面张力随浓度变化的关系曲线作切线，由其斜率求出某一指定浓度下溶液的表面吸附量。利用曲线作切线求微商的关键问题是如何准确地在曲线上作切线，常用的方法有两种：镜面法和平行线法。

① 镜面法　如图 1.6 所示。若需在 Q 点作切线，则可取一平而薄的镜子，使其边缘 AB 放在曲线的横断面上，绕 Q 转动，直至镜外曲线与镜像中曲线成一光滑曲线时，沿 AB 所画出的直线即为 Q 点的法线，作 Q 点法线的垂线即为该点的切线。

② 平行线法　如图 1.7 所示，在所选择的曲线上作两条平行线 AB、CD，作两线段中点的连线交曲线于 Q，过 Q 作与 AB、CD 之平行线即为 Q 点的切线。

（4）求经验方程　做出测量结果的函数关系的图形，以图形形式变换函数，使图形线性化，得到新函数 y 和新自变量 x 间的线性关系：

$$y = mx + b$$

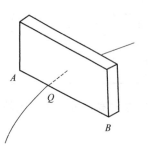
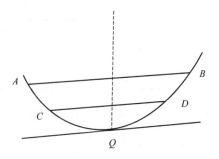

图 1.6　镜面法示意图　　　　　　图 1.7　平行线法示意图

以 y 对 x 作图，作一条尽可能连接各点的直线，由直线的斜率和截距求出线性方程中的 m 和 b，然后再换算成原函数和自变量，即得原函数的解析表达式。例如，溶液表面张力的测定实验中由兰格缪尔吸附等温式：

$$\Gamma = \Gamma_\infty \frac{Kc}{1+Kc}$$

求 Γ_∞ 时，若直接由 $\Gamma\text{-}c$ 曲线外推求 Γ_∞ 比较困难，则将上式进行变换为：

$$\frac{c}{\Gamma} = \frac{c}{\Gamma_\infty} + \frac{1}{K\Gamma_\infty}$$

使其直线化，即将 c/Γ 对 c 作图，由直线的斜率可求出 Γ_∞。

求取直线的斜率和截距可采用以下两种方法。

设直线方程为：

$$y = mx + b$$

欲求 m 和 b，一种方法是在直线上选取两点 (x_1, y_1)、(x_2, y_2)（选择时，两点不宜太近，以减小误差）代入上式，则：

$$m = \frac{y_2 - y_1}{x_2 - x_1}, \quad b = y_1 - mx_1 = y_2 - mx_2$$

另一种方法是延长直线与 y、x 轴相交，则直线与 x 轴的夹角 θ 的正切值 $\tan\theta$ 即为 b。上述两种方法中以采用前法最好。

(5) 求函数的极值或转折点　函数的极大、极小或转折点，在图形上表现直观且准确，因此在许多情况下都要应用它。例如，二元液系汽液平衡相图实验中，二元共沸物的恒沸点和组成的确定都常用作图法。

(6) 求面积计算相应的物理量（图解积分法）　设图形中的因变量是自变量导数的函数，求取曲线下自变量在一定取值范围内的面积即为因变量的定积分值。例如作 $x\text{-}y$ 曲线，可求得相应一定体积变化区间内曲线所包围的面积即为该过程所做的功。

1.4.2.2　作图技术

作图技术是利用图解法表达、处理实验数据取得优良结果的重要关键之一。作图时采用的工具主要有铅笔（HB 或 1H 为宜）、直尺和曲线板（应选用透明的）、曲线尺、圆规等。作图的一般步骤及规则如下。

(1) 坐标纸和比例尺的选择　通常所用的坐标纸有直角坐标纸、半对数和对数-对数坐标纸及三角坐标纸，在基础物理化学实验中最常用的是直角坐标纸。

用直角坐标纸作图时，应以自变量为横轴，以因变量为纵轴，坐标轴比例尺的选择一般遵循以下原则：

① 能表示出全部有效数字，以便从作图法求出的物理量的精确度与测量的精确度相适应；

② 图纸每小格对应的数值应便于迅速简便地读数，便于计算，如分度应为 1、2、5 或其倍数，避免 3、6、7、9 及其倍数；

③ 在满足上述条件下，考虑充分利用图纸的全部面积，若无必要，不必把坐标原点作为变量的零点，使图形布局匀称合理；

④ 若作的图是直线，则比例尺的选择应使其斜率接近于 1。

（2）画坐标轴　选定比例尺后，画上坐标轴，在轴旁注明该轴变量的名称、量纲及公共的乘方因子。在纵轴的左面和横轴的下面每隔一段距离写下该处变量应有的值，以便作图及读数。

（3）作代表点　把实验的测得值描点于图上，在点的周围画上圆圈、正方形、矩形或其他符号以区别各组的测量值。数据点周围的几何符号的面积大小应代表测量的精确度。若测量的精确度高，则圆圈的半径及矩形边长的半长度相应较小，反之则较大。

（4）连曲线　图纸上作好代表点后，按代表点的分布情况或作直线、或作曲线，表示代表点的平均变动情况。曲线的具体画法：先用淡铅笔轻轻地循各代表点的变动趋势，手描一条曲线（这条曲线不会十分平滑），然后用曲线板逐段凑合手描线的曲率，做出光滑的曲线。这里要特别注意各段接合处的连续性，做好这一点的关键是：①不要将曲线板上的曲边与手描线所有重合部分一次描完，一般只描半段或 2/3 段；②描线时用力要均匀，尤其在线段的起、终点时，更应注意用力适当。画线时，并不一定所有的数据点都在所绘的线上，但各点应在所绘曲线的两旁均匀分布，并使代表点与曲线间的距离的平方和为最小（即符合最小二乘法）。

（5）图名与说明　曲线作好后，还应在图上注上图名及比例尺和主要测量条件（包括温度、压力等）。最后写上姓名及实验日期。

值得注意的是，图是用形象来表达科学的语言，作图时应注意联系理论的基本原理，通常所作的曲线不应当有不能解释的间断点、突变点、自身交叉或其他不正常的特性。

1.4.3　数学方程式法

数学方程式法是将实验中各变量间的依赖关系用解析的形式表达出来。其优点是表达简单清晰、记录方便，也便于求微分、积分或内插值。由实验数据归纳出的解析式常称为经验方程式。经验方程式是客观规律的一种近似描述，是理论探讨的线索和依据，经验方程式中的系数往往与某一物理量相联系，因此采用该方法对实验数据进行处理也是非常必要的。

若实验中各变量间的依赖关系较简单时，寻找数学方程式中各常数项最方便的方法是将其直线化，即将函数 $y=f(x)$ 转换成线性函数：

$$y=mx+b$$

设法求出 m 和 b，若函数关系较为复杂，不能通过改换变量使原曲线直线化，可对原曲线进行非线性模型拟合，将其表达为 $y=f(x,b_1,b_2,\cdots,b_m)$ 的形式，应用计算机使用数值逼近的方法处理，求得各常数 b_1, b_2, b_3, $\cdots b_m$ 的值。

当实验各变量间的函数关系可设法表达为直线方程 $y=mx+b$ 的形式时，m 和 b 的求算可采用图解法、平均值法、最小二乘法三种方法获得。

1.4.3.1　平均值法

对线性方程 $y=mx+b$，原则上由两对变量 (x_1, y_1)、(x_2, y_2) 即可求出 m 和 b，但

由于实验测定中有误差存在，故这样处理的偏差较大。

平均值法的原理是正确的 m、b 值应使残差 u_i 之和为 0。u_i 是第 i 次测量的残差，其定义为：
$$u_i = mx_i + b - y_i$$

具体做法是将实验测得的数据 (x_1, y_1)、(x_2, y_2)、…、(x_i, y_i)、…、(x_n, y_n) 平均分两组 (x_1, y_1)、(x_2, y_2)、…、(x_k, y_k) 和 (x_{k+1}, y_{k+1})、(x_{k+2}, y_{k+2})、…、(x_n, y_n)。

通常 k 值大致为 n 的一半，代入式，得

$$\sum_{i=1}^{k} u_i = m \sum_{i=1}^{k} x_i + kb - \sum_{i=1}^{k} y_i = 0 \tag{1.34}$$

$$\sum_{i=k+1}^{n} u_i = m \sum_{i=k+1}^{n} x_i + (n-k)b - \sum_{i=k+1}^{n} y_i = 0 \tag{1.35}$$

联立上述二式，即可解出 m、b 的值。

该方法较图解法繁琐，但在有 6 个以上比较精密的数据时，结果比图解法好。

1.4.3.2 最小二乘法

最小二乘法是较为准确的方法，需要 7 组以上的数据，虽然处理较以上两法繁琐，但采用计算机处理，则十分简便，现已被广泛使用。

该方法的基本原理是在有限次数的测量中，其残差之和不一定为 0，但可以设想其最佳结果应能使其标准误差为最小，即 $\sum_{i=1}^{n} u_i^2$ 为最小。

令
$$S = \sum_{i=1}^{n} u_i^2 = \sum_{i=1}^{n} (mx_i + b - y_i)^2$$

则
$$S = m^2 \sum_{i=1}^{n} x_i^2 + 2bm \sum_{i=1}^{n} x_i - 2m \sum_{i=1}^{n} x_i y_i + nb^2$$

使 S 取极小值的必要条件是：

$$\left(\frac{\partial S}{\partial m}\right)_b = 0 = 2m \sum_{i=1}^{n} x_i^2 + 2b \sum_{i=1}^{n} x_i - 2 \sum_{i=1}^{n} y_i x_i \tag{1.36}$$

$$\left(\frac{\partial S}{\partial b}\right)_m = 0 = 2m \sum_{i=1}^{n} x_i + 2nb - 2 \sum_{i=1}^{n} y_i \tag{1.37}$$

联立上述二式，可解出

$$m = \frac{n \sum_{i=1}^{n} y_i x_i - \sum_{i=1}^{n} x_i \sum_{i=1}^{n} y_i}{n \sum_{i=1}^{n} x_1^2 - (\sum_{i=1}^{n} x_i)^2} \tag{1.38}$$

$$b = \frac{\sum_{i=1}^{n} x_i^2 \sum_{i=1}^{n} y_i - \sum_{i=1}^{n} x_i^2 y_i}{n \sum_{i=1}^{n} x_i^2 - (\sum_{i=1}^{n} x_i)^2} \tag{1.39}$$

使用最小二乘法时，可如表 1.7 那样将数据列成表格，在各栏末了算出加和结果，并把它代入方程式 (1.38) 和式 (1.39)，便可求得 m，b 的数值。

表 1.7　最小二乘法处理直线方程

x	y	x^2	xy
0.03	−3.01	0.0009	−0.0903
0.95	−0.97	0.9025	−0.9215
2.04	0.96	4.1616	1.9584
3.11	3.08	9.6721	9.5788
3.96	4.86	15.6816	19.2456
5.03	7.11	25.3009	35.7633
5.99	9.03	35.8801	54.0897
7.01	10.93	49.1401	76.6192
8.10	13.28	65.6100	107.5680
加和:36.22	45.27	206.3498	303.8113

$$m = \frac{n\sum_{i=1}^{n}y_i x_i - \sum_{i=1}^{n}x_i \sum_{i=1}^{n}y_i}{n\sum_{i=1}^{n}x_i^2 - (\sum_{i=1}^{n}x_i)^2} = \frac{9 \times 303.8113 - 36.22 \times 45.27}{9 \times 206.3498 - 36.22^2} = 2.008$$

$$b = \frac{\sum_{i=1}^{n}x_i^2 \sum_{i=1}^{n}y_i - \sum_{i=1}^{n}x_i^2 y_i}{n\sum_{i=1}^{n}x_i^2 - (\sum_{i=1}^{n}x_i)^2} = \frac{45.27 \times 206.3498 - 36.22 \times 303.8113}{9 \times 206.3498 - 36.22^2} = -3.049$$

例：设有直线方程 $y = mx + b$，x 和 y 的数值如下：

x	3.36	3.30	3.25	3.20	3.14
y	−3.64	−3.15	−2.72	−2.29	−1.88

将上述数据依次代入式(1.38) 和式(1.39)，得下列 5 个方程：

$b + 3.36m = -3.64$　①　　$b + 3.30m = -3.15$　②

$b + 3.25m = -2.72$　③　　$b + 3.20m = -2.29$　④

$b + 3.14m = -1.88$　⑤

将①～③为一组，④～⑤为一组代入式(1.36) 和 (1.37)，得

$$\begin{cases} 3b + 9.91m + 9.51 = 0 \\ 2b + 6.34m + 4.17 = 0 \end{cases}$$

解之，得 $m = -8.14$　　$b = 23.7$

$$\sum_{i=1}^{5} u_i = 0.136$$

采用最小二乘法，得

$m = -8.10$　　$b = 23.6$

$$\sum_{i=1}^{5} u_i = 0.08$$

可知最小二乘法处理时，残差更小。

第2章 基础实验

实验2.1 恒温槽的性能调节及液体黏度测定

一、实验目的

1. 了解恒温槽的结构及恒温原理；掌握恒温槽的使用方法；测绘恒温槽的灵敏度曲线并计算灵敏度。
2. 了解黏度的概念；掌握用奥氏黏度计测定液体黏度的方法。

二、实验原理

1. 恒温槽及其控温原理

物质的许多物理化学性质，如饱和蒸气压、黏度、电导以及化学反应的平衡常数与速率常数等，都与温度密切相关。因此，大多数物理化学常数的测量需要在恒温条件下进行。恒温槽是实验中常用的一种以液体为介质的恒温装置。本实验采用的恒温槽以水为介质，用于室温以上、80℃以下的温度范围，其构造见图2.1。

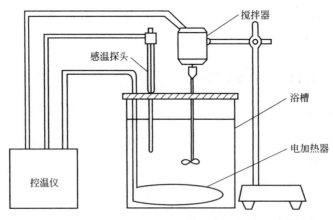

图2.1 恒温槽

图2.1所示的浴槽是敞口透明圆玻璃缸，便于观察浸入浴槽中的器皿。感温探头为温差热电偶，将介质的温度信息转换为电信号。

恒温槽的控温原理为：将控温仪的设定温度调至所需温度（如25℃），此时感温探头将探测的介质温度（应低于设定值）与设定温度相比较，如低于设定温度，则温控仪命令介质中的加热器工作（此时温控仪上加热指示灯亮），使介质的温度升高，搅拌器保持介质温度均匀。当介质温度达到设定值时，温控仪指令停止加热（此时加热指示灯灭，恒温指示灯亮）。随着介质向外散热，温度会逐渐低于设定值，温控仪指令加热器工作，如此反复进行控制，自动保持介质温度的恒定。

由于传热、感温都需要一定的时间，这会使补充热量与温度的升降之间存在滞后，导致恒温槽内介质的温度在一定范围内波动。这种温度随时间而变化的曲线，称为恒温槽的灵敏度曲线。在灵敏度曲线上，最高温度与最低温度之差的一半，称为恒温槽的灵敏度。灵敏度是衡量一台恒温槽控制精度的物理量。物理化学实验使用的恒温槽，通常要求其灵敏度在 ± 0.1℃之内。恒温介质的准确温度由槽内悬挂的精密温度计显示，温控仪上的温度显示不能作为准确值进行读数，仅作为大致参考。

2. 黏度的测定原理

黏度是流体的一种重要性质。它反映了流体流动时由于各点速度不同而产生的剪切应力大小。许多流体在流动时，任一微分体积单元上剪切应力与垂直于流体方向的速度梯度成正比，这种流体称为牛顿（Newton）型流体。几乎所有的气体和许多简单的液体都是牛顿型流体。对牛顿型流体，剪切力 F（即流动时的内摩擦力）与流速梯度 du/dy 及接触面积 A 之间符合下述关系：

$$F = -\eta A du/dy \tag{2.1}$$

式中负号表示剪切力的方向与流动方向相反。式中比例系数 η 称为绝对黏度（简称黏度），其物理意义为：在流体中两个相距单位长度的具有单位面积的流体层，以单位流速发生相对运动时所需剪切力的大小。当 $A=1m^2$，$du/dy=1s^{-1}$，F 恰好为 1N 时，绝对黏度 η 值为 $1N \cdot s \cdot m^{-2}$（1牛顿·秒·米$^{-2}$），即 $1Pa \cdot s$（1帕·秒）。

本实验用于测定牛顿型液体的黏度。测定液体绝对黏度的方法，主要分成三类：
① 测定液体在毛细管中流过的时间——毛细管法；
② 测定圆球在液体中落下的时间——落球法；
③ 测定液体在同心轴圆柱筒体之间对筒体相对转动的影响——转筒法。

毛细管法根据使用的黏度计的不同又各不相同。下面介绍奥氏（Ostwald）黏度计（见图2.2）的使用。液体在毛细管黏度计中因重力作用而流动，流动特性遵守泊肃叶（Poiseuille）方程：

$$\eta = \frac{\pi r^4 g h d t}{8 V l} \tag{2.2}$$

式中，r 为毛细管内半径；l 为毛细管长度；g 为重力加速度；h 为黏度计两管内平均液柱差；d 为液体密度；V 为从毛细管中流过的液体的体积，即 A、B 刻度之间的管内体积；t 为体积为 V 的液体流过毛细管所需的时间。对于一支黏度计，r、V、l 为定值。当液体在毛细管内流动时，黏度计两管内液柱差是变化的。

式中 h 代表黏度计内一侧液面从刻度 A 流至刻度 B 期间，与实际流动情况等效的两侧内平均液柱高度差。此 h 值在黏度计几何形状一定、黏度计中液体体积一定情况下为定值。因此，在加入黏度计的液体体积一定的条件下，泊肃叶方程可以改写为：

$$\eta = K d t \tag{2.3}$$

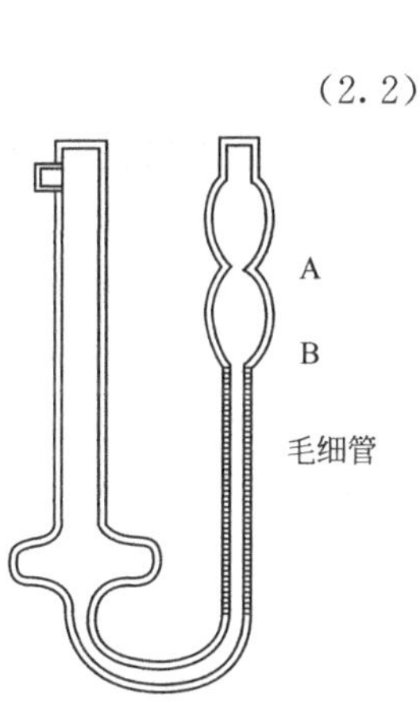

图2.2 奥氏（Ostwald）黏度计

$$K = \frac{\pi r^4 g h}{8Vl} \tag{2.4}$$

式中，K 称为黏度计常数。利用已知黏度的液体（如水或甘油水溶液），测定其密度及流过时间，即可算出该毛细管黏度计常数 K。此 K 值受温度影响，但影响很小。

对于一般液体，温度越高，黏度越小，温度与黏度的关系可用下列经验公式表示：

$$\ln \eta = A/T + B \tag{2.5}$$

式中，A、B 为经验常数，其值因液体而异。由于黏度受温度影响较大，所以测定黏度必须在恒温条件下进行。

三、仪器与试剂

图 2.1 所示恒温槽装置一套；温控仪一台；贝克曼温度计；1/10 温度计；秒表；奥氏黏度计；移液管 2 支；滴管 2 支；乙醇（AR）；小烧杯；甘油水溶液。

四、实验步骤

1. 连接装置与温度设定（30.0℃）

按图 2.1 接好恒温槽线路。打开温控仪开关，将圆形刻度盘旋钮调至设定温度 30℃。若被控介质温度低于设定温度，则温控仪上加热指示灯亮，恒温槽中加热器开始加热（可观察到加热管外表面有微小气泡逸出），温度开始上升，升至设定温度时，恒温指示灯亮，温控仪进行自动恒温。开动搅拌器，选择适当转速进行搅拌。

温控仪上的设定旋钮因经常使用会产生误差，其校正方法如下：

先将设定旋钮调至 25℃（比正式设定值低 3~5℃，但比介质温度要高），加热指示灯亮，至停止加热（恒温指示灯亮）时，恒温槽内精密温度计上的读数应为 25℃，如有误差，则此值为设定旋钮之误差，据此进行校对。

由于这类恒温槽是靠自然散热而降温的，而且浴槽的热容较大，温度高于指定温度后，降温很慢。故进行上述调节时要防止温度过高。

2. 恒温槽灵敏度的测定

待恒温槽已调节到 30℃后，观察精密温度计上的读数，利用秒表，每隔 1min 记录一次温度计的读数。测定 30~40min，温度变化范围要求在 ±0.15℃。

按同样步骤测定 35℃时恒温槽灵敏度。

3. 液体黏度的测定

在洁净干燥的毛细管黏度计粗管端套一软木塞，用烧瓶夹夹住软木塞。将黏度计竖直浸在恒温槽中，使槽内水面没过刻线 A。将烧瓶夹固定在支架上。为保证黏度计的毛细管部分垂直，需用一铅垂线从两个相互垂直的方向进行检查。

用移液管取 10~12mL 乙醇或甘油水溶液，从粗管口放入黏度计。在细管口接一段乳胶管。恒温 10min 后，从乳胶管口用洗耳球将溶液的液面吸到 A 刻线以上，让其自然流下（如为三管黏度计，则用手指堵住无毛细管的细支管，从中间细管上吸液体）。在液面恰好通过 A 刻线时，启动秒表。在液面恰好通过 B 刻线时，停止秒表，记下时间（秒）。重复操作 3 次，误差应小于 1s。取 3 次时间的平均值，作为甘油水溶液流出的时间。

取下黏度计，弃去溶液，用蒸馏水仔细洗 3 次。再用移液管取同量蒸馏水，按上述方法，测定蒸馏水流出时间，测 3 次，取平均值。记下黏度计上的号码。

五、注意事项

1. 搅拌转速应适度，不宜过快，能使介质温度均匀即可。
2. 调节设定温度时宁低勿高，再缓慢调升至设定值。
3. 黏度计安装时其毛细管部分务必垂直。
4. 贝克曼温度计和黏度计易损坏，操作时应特别小心。
5. 夏天室温高，可选择35℃恒温并进行测定。

六、数据处理

1. 实验数据记录于表2.1和表2.2中。计算待测液体在所测温度下的黏度，并与文献值比较。
2. 绘出恒温槽的灵敏度曲线，计算恒温槽的灵敏度［灵敏度 $\Delta T=(T_{max}-T_{min})\div 2$］。

表 2.1　实验数据记录

室温：_____　气压：_____
d（甘油水溶液）：　　　kg·m^{-3}；d（蒸馏水）：　　　kg·m^{-3}
甘油水溶液黏度的测定：

样品 \ 实验编号 时间(t)/s	1	2	3
蒸馏水			
甘油水溶液			

表 2.2　恒温槽灵敏度的测定

时间/min	1	2	3	4	5	6	7	8	9	10
温度/℃										
时间/min	11	12	13	14	15	16	17	18	19	20
温度/℃										
时间/min	21	22	23	24	25	26	27	28	29	30
温度/℃										

七、思考题

1. 根据恒温槽温度波动的原因及规律，可采用哪些措施提高恒温槽控温精密度？
2. 黏度计放在恒温槽内为什么必须垂直？它在泊肃叶方程中通过哪个物理量来影响黏度测量值？
3. 为何注入黏度计的试液与标准液的体积必须相同？

Experiment 2.1　Performance of Thermostatic Bath and Determination of Liquid Viscosity

Objectives

1. To know the structure of the thermostatic bath and to understand the basic principle

of it.

2. To learn how to use the thermostatic bath.

3. To determine the sensitivity of the thermostatic bath and to draw the sensitivity curves of it.

4. To understand the concept of viscosity and to learn how to determine the liquid viscosity by Ostwald Viscometer.

Principle

1. The thermostatic bath and its temperature control principle

Many physical and chemical properties of the material are closely related to temperature, such as the saturation vapor pressure, viscosity, conductivity, and chemical reaction equilibrium constant and rate constants, etc. Therefore, most of the physical and chemical constants are measured at a constant temperature. The thermostatic bath used a liquid medium is the thermostatic apparatus as shown in Fig. 2.3. In this study, the thermostatic bath uses water as the medium, and the range of temperature is from the room temperature to 80 ℃.

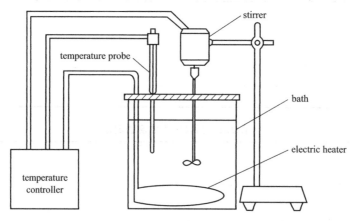

Fig. 2.3 The thermostatic bath

Temperature probe is thermocouple which converts the medium temperature information to electrical signals. Temperature sensing and heat transfer will take some time and lead to some lag between the supplementary heat and the temperature lift, resulting in the temperature of the medium fluctuates within a certain range. This curve, temperature against time, is called the sensitivity curve of the thermostatic bath.

In the sensitivity curve, half of the difference value of the maximum temperature and minimum temperature is called the sensitivity of the thermostatic bath. Exact temperature of the heated medium should be displayed by the microthermometer, while temperature displayed on the temperature controller should be taken as reference only.

2. The measurement principle of the liquid viscosity

Viscosity is an important property of fluid. It reflects the size of the generated shear stress. The shear stress of a differential volume unit is proportional to the perpendicular direction of fluid velocity gradient when fluid flows, this type of fluid is called Newtonian liquid. The shear stress obeys the formula (1)

$$F = -\eta A du/dy \tag{2.1}$$

Where negative sign indicates the direction of the shear stress and flow is in the opposite direction. Where the scale factor η is called the absolute viscosity (Referred to as the viscosity).

This experimental determination of the viscosity is the type of Newtonian liquid. There are three kinds of methods to measure the viscosity of a liquid:

(1) To measure the time required for the liquid to flow through a capillary tube with a capillary viscometer;

(2) To measure the falling speed of a ball in the liquid with a ball viscometer;

(3) To measure the relative rotation between the liquid and a coaxial cylinder with a rotation viscometer.

Capillary method is different according to the viscometer. We use Ostwald viscosimeter as shown in Fig. 2.4. When the liquid flows through the capillary viscometer because of gravitation, it obeys the Poiseuille law:

$$\eta = \frac{\pi r^4 ghdt}{8Vl} \tag{2.2}$$

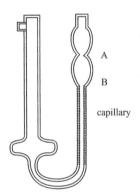

Fig. 2.4 Ostwald viscosimeter

Where r is the capillary radius, l is the capillary length, g is the gravity acceleration, h is the average height of liquid column passing through the capillary tube, d is the liquid density, V is the volume of liquid passing through the capillary tube, t is the flow time.

The value of r and V and l is definite for a given viscometer. The value of h is definite when the viscometer is given and the volume of liquid in the viscometer is invariable. Therefore, when the liquid volume joined into the viscosity is invariable, the Poiseuille equation can be written as:

$$\eta = Kdt \tag{2.3}$$

$$K = \frac{\pi r^4 gh}{8Vl} \tag{2.4}$$

Where K is the viscometer constant. Calculate constant K of the capillary viscometer by measuring the density and flow time of the liquid where the viscosity is known. The value of K is affected a little by temperature. For general liquid, when the temperature is higher, the viscosity is smaller as the following empirical equation.

$$\ln\eta = A/T + B \tag{2.5}$$

Apparatus and Reagent

Thermostatic water bath, 1; temperature control unit, 1; Beckman thermometer, 1; 1/10 thermometer, 1; second chronograph, 1; Ostwald viscosimeter, 1; 10 mL pipette, 2; dropper, 2; ethanol (AR); beaker; glycerol solution.

Procedure

1. Install the equipments and set temperature for the temperature control unit. Set the

thermostatic water bath's temperature to a specified temperature (30.0℃). Turn on the power of the thermostatic water bath and stirrer. We must prevent temperature from being too high.

2. Measure the sensitivity of thermostatic water bath. When the bath is adjusted to 30℃, observe the precision thermometer and record the readings of the thermometer every one minute. The measurement time is about 30 ~ 40 minutes.

3. Measure the liquid viscosity.

Clean the viscometer and clamp the viscometer vertically on an iron stand, and place it in the thermostatic water bath, and make water surface higher than line 'A. Pipet 10.0mL ethanol or glycerol solution carefully into the viscometer. After 10 minutes' constant temperature, suck up the solution by the rubber suction bulb above line A, then let the solution fall along the capillary tube naturally. Start the stopwatch exactly as the liquid meniscus passes line A, and stop it just as the meniscus passes line B. Record the collapsed time for the liquid to fall from A to B. Repeat this measurement for three times. The error should be within 0.2 s. Average the flow time to obtain the t value.

Take the viscometer out and clean it with distilled water. Measure the t value through distilled water with the same amount of glycerol solution by the above method. Write down the number of the viscometer.

Notes

1. Stirring speed should be moderate and make sure the medium temperature is even.

2. Adjust the set temperature to a low value, then slowly raised the temperature to setting value.

3. The viscometer must be installed vertically.

4. Operation should be especially careful not to damage the Beckman thermometer and viscometer.

5. In summer we can select 35℃ as the constant temperature.

Data Analysis

1. Calculate the viscosity of the liquid sample in the measured temperature, and compare it with literature values.

2. Draw the sensitivity curve of the thermostatic bath and calculate the sensitivity of it. [Sensitivity $\Delta T = (T_{max} - T_{min}) \div 2$]

Questions and Further Thoughts

1. According to the causes and laws of the thermostatic bath temperature fluctuations, what measurments can be used to enhance temperature control precision of the thermostatic bath?

2. Why the viscometer must be installed vertically? Through which physical parameters it affect the viscosity measurement in the Poiseuille equation?

3. Why the solution injected into the viscometer must be the same volume of standard solution?

实验 2.2　易挥发物质摩尔质量的测定

一、实验目的

用维克托-梅耶（Victor Meyer）法测定乙醇的摩尔质量，要求掌握质量、温度、压力、体积测量的基本操作。

二、实验原理

在温度不太低、压力不太高的条件下，可近似地把实际气体看作理想气体，其状态方程为：

$$M=\frac{mRT}{pV} \qquad (2.6)$$

式中，p、V、T、m 和 M 分别为气体的压力、体积、温度、质量和摩尔质量；R 为通用气体常数。

将一定质量的易挥发液态物质在保持温度（通常较该物质沸点高 20～30℃）及压力（通常为大气压力）恒定的容器底部汽化，此蒸气将把容器中与该蒸气同温、同压力和同体积的空气排挤出来，排出的空气在常温常压下不会液化，容易测出它的 p、V、T，从而可算出其物质的量，其值与液体蒸气物质的量相等。已知液体的质量 m，即可算出被测物质的摩尔质量。

物质在汽化时需防止蒸气扩散至汽化管上部低温区冷凝而使排出空气的体积减少。同理，实验前汽化管中不应含凝结的蒸气。

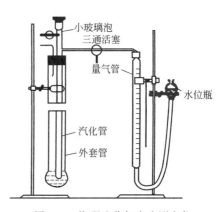

图 2.5　梅耶法蒸气密度测定仪

三、仪器与试剂

1. 装置如图 2.5 所示；0～100℃ 温度计 1 支；小玻璃泡数个；电炉 1 个；分析天平 1 台。

2. 无水乙醇（AR）；酒精灯。

四、实验步骤

1. 按图 2.3 装好测定仪。将三通活塞通大气，接通电炉电源，将外管中的盐水加热至沸（水中加少量沸石）。

2. 取一干净小玻泡在分析天平上精确称量（准确至 0.0002g）。将小玻泡用酒精灯加热后，迅速将玻泡的毛细管端插入待测液体中，玻泡冷却后，液体即被吸入（若待测液体为无水乙醇，则吸入的质量为 0.10～0.16g；若待测液体为乙酸乙酯，吸入的质量为 0.1～0.2g。过多过少都不适宜）。大致称量使质量适合后，于酒精灯上熔封玻泡毛细管尖端。然后重新准确称量。两次质量之差即为待测物的质量。

3. 将小玻泡小心地放入汽化管上部的玻璃棒上，塞紧管口塞子，检查体系是否漏气，即转动三通活塞，使汽化管与量气管相连，提高或降低水位瓶后观察一段时间，若量气管液面上升或下降一定程度后不再变化,则表明不漏气;若量气管液面不断上升或下降,则表明漏气,应重新检查装置。

4. 确认装置不漏气后,加热使外管水沸腾约 5min,检查汽化管内温度是否恒定:转动三通活塞,使量气管与大气相通。将水位瓶慢慢往上提,使量气管与水位瓶的液面平齐,再旋转三通活塞,使汽化管和量气管相连,若管内液面并不上下移动,表明温度已达稳定。

5. 汽化管内温度恒定后,使三通活塞与大气相通,提高水位瓶,使量气管中水面升至顶点附近(注意勿使液体进入横向连接管),然后旋转三通活塞,使内管与量气管相连,保持量气管与水位瓶液面等高,记下此时量气管的读数作为初始体积。轻轻拉长玻璃棒套(注意勿使装置漏气),使玻璃棒外移,小玻泡失去支撑落入汽化管的底部而破碎,小泡内部的液体立即汽化而将内管上部空气排入量气管中。移动水位瓶,保持水位瓶与量气管两个水平面等高,至液面不再下降,记下该处读数作为终了体积;并记录靠近量气管的温度及大气压力(表 2.3)。

表 2.3 实验数据记录

室温:_____ 气压_____

实验编号	玻泡质量 m/g	玻泡+待测液质量 m/g	待测液质量 m/g	量气管初读数 V/cm³	量气管末读数 V/cm³	排出空气体积 V/cm³	大气压力 p/Pa	空气温度 t/℃
1								
2								

6. 转动三通活塞通大气,停止加热。将内管取出,倒出碎玻璃,干燥汽化管,赶出残存的液体。

重复以上步骤,做一次平行实验。

五、数据处理

1. 查出在量气管的温度下,水的蒸气压 $p^*(H_2O)$,求出量气管内空气(不含水蒸气)的分压 p:

$$p = p_{大气} - p^*_{H_2O}$$

2. 根据所测得的 V、T、m 等数据,利用理想气体状态方程,求出所测物质的摩尔质量。

3. 将实验算得的 M 与计量式的摩尔质量比较,求出百分误差。

六、思考题

1. 汽化管与量气管温度不同,为什么可以在量气管中测量被排出气体的 V、T 来计算汽化管内待测物的摩尔质量?

2. 样品太多或太少有什么不好?

3. 每次实验完毕为什么要把内管中样品排净?

4. 汽化管中气体上下存在温度梯度,会影响实验结果吗?

5. 读量气管读数量为什么水准瓶要与量气管内水面平齐?

Experiment 2.2 Determination of Molar Mass by Evaporization

Objectives

Measuring the molecular weight of anhydrous ethanol by Victor-Meyer method; mastering the basic operation measuring quality, temperature, pressure, volume.

Theory/Principle

In the condition that temperature is not too low and pressure is not too high, any actual gas is similar to an ideal gas, and its equation of state is as follows:

$$M = \frac{mRT}{pV} \tag{2.6}$$

Where p, V, T, M and m stand for gas pressure, volume, temperature, molecular weight and quality respectively; R is a common gas constant.

A certain quality of volatile liquid will be evaporated in the bottom of the container under the constant temperature (usually 20~30 ℃ higher than the boiling point of the measured material) and pressure (usually atmospheric pressure). This vapor will drive off the air at the same temperature, pressure and volume. The air under normal temperature and pressure can not be liquefied, and its T, p, V are easily measured, thus the molecular weight can be figured out.

Apparatus and Chemicals Required

A thermometer: 0~100 ℃; several small glass bubbles; a furnace; an analytical balance; anhydrous ethanol; an alcohol lamp.

Procedure

1. Installing the instrument as shown in Fig. 2.6. It must be interlinked with the atmosphere. Putting on the electric power to heat. Making the salt water in the jacket boiling.

2. Weighing a small glass bubble with an analytical balance (accurate to 0.0002g). Heating the small glass bubble with an alcohol lamp, then inserting it into anhydrous ethanol quickly. When the small glass bubble cools down, anhydrous ethanol will be absorbed into it. The mass of the anhydrous ethanol absorbed is required from 0.10~0.16g. The capillary top of the glass bubble will be sealed through melting when the mass meets requirement, then weighing the total mass accurately.

3. Putting the small glass bubble containing anhydrous ethanol carefully on the glass rod upside within the gasification tube, covering the tube tightly with a stopper, then checking wheth-

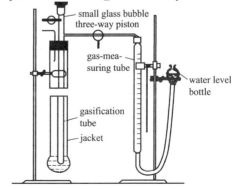

Fig. 2.6 The instrument of steam density by Meyer method

er the system leaks.

4. Heating the water in the jacket if there is no leakage, checking whether the temperature within the gas-measuring tube has been stable after boiling about 5 min. If liquid in the gas-measuring tube does not move up and down, it means the temperature has been stable.

5. Regulating water level in the gas-measuring tube to select the beginning volume when the temperature is stable. The beginning volume may not be too big, at best 10 mL. After recording the beginning volume and the temperature of the gas-measuring tube, pulling the glass rod lightly to make the small glass bubble drop down. If the small glass bubble fall and break, anhydrous ethanol liquid will be vaporized, the volume will increase. Recording the final volume when the volume has been stable, and recording the atmospheric pressure.

6. Rotating three-way piston to link with atmosphere, and stopping heating. Removing the gasification tube, pouring broken glass, and drying the gasification tube with Drying oven.

Observations and Measurements

Table 2.4   room temperature: _____   atmospheric pressure: _____   air temperature: _____

mass of glass bubble $(m)/g$	mass of glass bubble and anhydrous ethanol $(m)/g$	mass of anhydrous ethanol $(m)/g$
the beginning volume $(V)/cm^3$	the final volume $(V)/cm^3$	volume of anhydrous ethanol vapor $(V)/cm^3$

Calculations

1. Checking the saturated vapor pressure of water at the air temperature in the gas-measuring tube, the air pressure in the gas-measuring tube is as follows:

$$p = p_{大气} - p^*_{H_2O}$$

2. Figuring out the molecular weight of anhydrous ethanol by the formula (2.6).

3. Comparing the molecular weight of anhydrous ethanol obtained by the experiment with the true molecular weight of anhydrous ethanol, and calculating the percentage error.

Notes

1. Preventing the vapor of anhydrous ethanol condensing on the top of the gasification tube.

2. There must not be any liquid in the gasification tube before the experiment.

3. The water in the gas-measuring tube must be at the same height as the water in the water level bottle when reading the volume.

Questions/Exercises

1. The temperature of gasification tube is different from the temperature of gas-measuring tube. Why can the molecular weight of anhydrous ethanol be figured out with the air temperature and volume in the gas-measuring tube?

2. Why can't the sample be too much or too little?

3. Why should the sample in the tube be cleaned out after the experiment?

4. There is the temperature gradient in the gasification tube. Will the experimental result be affected?

5. Why must the water in the gas-measuring tube be at the same height as the water in the water level bottle when reading the volume?

实验 2.3 溶解热的测定

A. 测温量热法

一、实验目的

用量热法测定 KNO_3 在水中的溶解热，掌握测温量热的基本原理和测量方法，学会作图外推法求真实温差。

二、实验原理

溶解热是物质溶解于溶剂时产生的热效应，其值受温度、压力、溶质和溶剂量的影响。一般将其分为微分溶解热和积分溶解热两种，本实验测定积分溶解热，其定义为：在一定的温度、压力下，1mol 溶质溶解于一定量溶剂中形成一定浓度溶液时的热效应。

盐在水中的溶解过程主要是盐的晶格的破坏和离子的溶剂化，则盐的溶解热是破坏盐的晶格的热效应和离子溶剂化热效应的总和。

测量热效应的量热计一般可分为两类：一类是等温量热计，其本身温度在量热过程中始终不变，测定物理量为体积的变化；另一类是经常采用的测温量热计，通过测定量热过程中温度的变化进行量热。本实验采用的测温量热计，是一个包括量热容器、搅拌器和温度计的量热系统（如图 2.7 所示）。

当实验在恒压无非体积功的绝热系统中进行时，系统的总焓保持不变，可列出如下的热平衡式：

$$\Delta_{sol}H_m + \Delta H_1 + \Delta H_2 + \Delta H_k = 0 \quad (2.7)$$

式中，$\Delta_{sol}H_m$ 为盐在溶液温度及浓度下的积分溶解热；ΔH_1、ΔH_2、ΔH_k 分别为由于温度变化引起的溶质、溶剂和量热计的热效应，由式

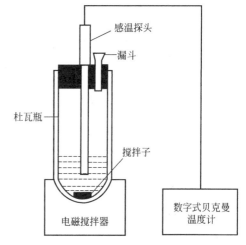

图 2.7 溶解热测定装置

(2.7) 可得：

$$\Delta_{sol}H_m = -(\Delta H_1 + \Delta H_2 + \Delta H_k) = -(W_1 C_1 + W_2 C_2 + K)\frac{M_1}{W_1}\Delta T \quad (2.8)$$

式中，W_1 和 W_2 分别为溶质和溶剂的质量，g；C_1 和 C_2 分别为溶质和溶剂的比热容；K 为量热计的比热容；M_1 为溶质摩尔质量；ΔT 为溶解过程的真实温差。

用已知溶解热的物质测得 W_1、W_2 和 ΔT，按式(2.8)计算量热计的热容 K；然后测定 KNO_3 溶解过程的 W_1、W_2 和 ΔT，即可按式(2.8)算出其 $\Delta_{sol}H_m$。

三、仪器与试剂

量热计一套；电磁搅拌器一台；500mL 容量瓶一个；秒表一块；数字贝克曼温度计一台；1/10 温度计 1 支（0~50℃）；KNO_3（AR）；KCl（AR）。

四、实验步骤

1. 量热计热容的测定

本实验采用已知氯化钾在水中的溶解热来标定量热计热容 K（不同温度下 1mol KCl 溶于 200mL 水中的积分溶解热见附录附表8）。先在干净的量热计中装入 500mL 蒸馏水，插入数字式贝克曼温度计感温探头。打开电磁搅拌器以一定的速度搅拌（速度不宜过快），到温度基本稳定后，每分钟读数一次，连续 8 次后，打开盖子，迅速倒入称好的氯化钾（其重量按 1mol KCl：200mol H_2O 计算，称准至 0.01g）盖好量热计盖子，并仍按每分钟读数一次，至温度不再下降后再读 8 次即可停止。关上搅拌器，然后用普通温度计测出量热计中溶液的温度。

2. 硝酸钾溶解热的测定

用硝酸钾（其重量按 1mol KNO_3：400mol H_2O 计算）代替氯化钾重复以上操作。实验数据记录于表 2.5 和表 2.6 中。

表 2.5 实验数据记录（一）

项目	加入 KCl	加入 KNO_3
盐的质量 m/g		
量热计水的体积 V/cm³		
溶液温度 T/℃		

表 2.6 实验数据记录（二）

加入 KCl	时间 t/min		
	温度 T/℃		
加入 KNO_3	时间 t/min		
	温度 T/℃		

五、数据处理

1. 绘制温度-时间曲线，求真实温差 ΔT。由于量热计并不是严格的绝热系统，为消除盐溶解过程中系统与环境有微小的热交换造成的影响，可以用作图外推法求真实温差，方法如下。

(1) 作温度-时间曲线如图 2.8 所示，A 点相当于热效应开始点，B 点相当于热效应终点，AB 称为主期，PA 为前期，BQ 为后期。

(2) 取 AB 两点间的平均温度为中点交曲线于 D 点。

(3) 过 D 点作一垂线分别交 PA 和 BQ 的延长线于 E 和 F 两点，则 EF 线段表示真实温差 ΔT。

2. 计算量热计热容 K。KCl(s) 及 KNO$_3$(s) 在 20℃ 附近的比热容分别为 0.669 J·g^{-1}·K^{-1} 和 0.805 J·g^{-1}·K^{-1}。KCl 的溶解热可查第 5 章 5.8。

3. 计算硝酸钾在溶液温度下的溶解热。

将上述结果列于表 2.7。

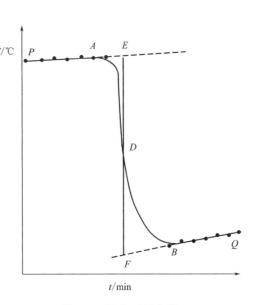

图 2.8　温度-时间曲线

表 2.7　量热计热容测定

项目	量热计热容的测定	KNO$_3$ 溶解热的测定
溶解温度 T/℃		
真实温差 ΔT/℃		
水的比热容 C_2/J·g^{-1}·K^{-1}		
量热计热容 K/J·K^{-1}		
KNO$_3$ 溶解热 $\Delta_{sol}H_m$/J·mol^{-1}		

六、思考题

1. 为什么只做膨胀功的绝热体系在等压过程中总焓不变？

2. 温度和浓度对溶解热有无影响？如何从实验温度下的溶解热计算其他温度下的溶解热？

B. 电热补偿法

一、实验目的

掌握量热装置的基本组合及电热补偿法测定热效应的基本原理；用电热补偿法测定 KNO$_3$ 或 KCl 在不同浓度水溶液中的积分溶解热；用作图法求 KNO$_3$ 或 KCl 在水中的微分冲淡热、积分冲淡热和微分溶解热。

二、实验原理

1. 基本概念

物质溶解于溶液中一般伴随有热效应发生，热效应的大小取决于溶剂、溶质的物质本性和它们的相对量。在热化学中，关于溶解过程的热效应有下列几个基本概念。

(1) 溶解热　在恒温恒压下，n_2 mol 溶质溶于 n_1 mol 溶剂（或溶于某浓度溶液中）产生的热效应，用 Q 表示，溶解热可分为积分（或称变浓）溶解热和微分（或称定浓）溶解热。

① 积分溶解热　在恒温恒压下，1mol 溶质溶于 n_0 mol 溶剂中产生的热效应，用 Q_s 表示。

② 微分溶解热　在恒温恒压下，1mol 溶质于某一确定浓度的无限量的溶液中产生的热效应，以 $\left(\dfrac{\partial Q}{\partial n_2}\right)_{T,p,n_1}$ 表示，简写为 $\left(\dfrac{\partial Q}{\partial n_2}\right)_{n_1}$。

（2）冲淡热　在恒温恒压下，1mol 溶剂加到某浓度的溶液中使之冲淡所产生的热效应。

① 积分冲淡热　在恒温恒压下，把原含 1mol 溶质及 n_{01} mol 溶剂的溶液冲淡到含溶剂 n_{02} mol 时的热效应，即为某两浓度溶液的积分溶解热之差，用 Q_d 表示。

② 微分冲淡热　在恒温恒压下，1mol 溶剂加入某一确定浓度的无限量的溶液中产生的热效应，以 $\left(\dfrac{\partial Q}{\partial n_1}\right)_{T,p,n_2}$ 表示，简写为 $\left(\dfrac{\partial Q}{\partial n_1}\right)_{n_2}$。

2. 电热补偿法

本实验采用电热补偿法测定热效应。积分溶解热 Q_s 可由实验直接测定，其他三种热效应通过 Q_s-n_0 曲线求得。

以 Q_s 对 n_0 作图，可得图 2.6 的曲线。

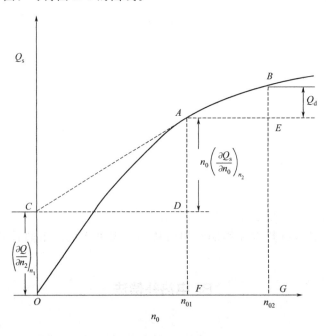

图 2.9　Q_s-n_0 关系图

在图 2.9 中，AF 与 BG 分别为将 1mol 溶质溶于 n_{01} mol 和 n_{02} mol 溶剂时的积分溶解热 Q_s，BE 表示在含有 1mol 溶质的溶液中加入溶剂，使溶剂量由 n_{01} mol 增加到 n_{02} mol 过程的积分冲淡热 Q_d。计算公式：$Q_d=(Q_s)_{n_{02}}-(Q_s)_{n_{01}}=BG-EG$。

图 2.9 中 A 点的切线斜率等于该浓度的微分冲淡热 $\left(\dfrac{\partial Q}{\partial n_1}\right)_{n_2}$，其计算公式如下：

$$\left(\frac{\partial Q}{\partial n_1}\right)_{n_2}=\left(\frac{\partial Q/n_2}{\partial n_1/n_2}\right)_{n_2}=\left(\frac{\partial Q_s}{\partial n_0}\right)_{n_2}=\frac{AD}{CD} \tag{2.9}$$

切线在纵轴上的截距等于该浓度的微分溶解热 $\left(\dfrac{\partial Q}{\partial n_2}\right)_{n_1}$，计算公式如下：

$$\left(\frac{\partial Q}{\partial n_2}\right)_{n_1} = \left(\frac{\partial n_2 Q_s}{\partial n_2}\right)_{n_1} = Q_s - n_0 \left(\frac{\partial Q_s}{\partial n_0}\right)_{n_2} = OC \tag{2.10}$$

3. 实验方法

本实验采用绝热式测温量热计，它是一个包括量热器、搅拌器、电加热器和测温部件等的量热系统。装置及电路图如图 2.10 所示。因本实验测定的样品 KNO_3 在水中的溶解是一个吸热过程，可用电热补偿法测定，即先测定体系的起始温度 T，溶解过程中体系的温度随吸热反应进行而降低，用电加热法使体系升温至起始温度，根据消耗电能求出热效应 Q。

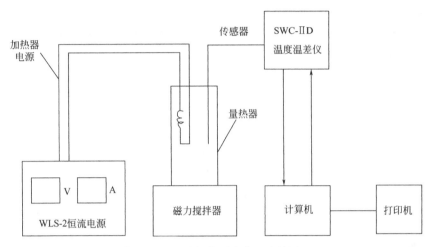

图 2.10 溶解热测定装置连接图

$$Q = I^2 Rt = UIt \tag{2.11}$$

式中，I 为通过电阻为 R 的电加热器的电流强度，A；U 为电阻丝两端的所加电压，V；t 为通电时间，s。进而可求得积分溶解热 Q_s：

$$Q_s = \frac{Q}{n_2} \tag{2.12}$$

$$Q = IUt_{累} \tag{2.13}$$

$$n_2 = \frac{W_{累}}{M_2} \tag{2.14}$$

式中，$t_{累}$ 为各次加入 KNO_3 或 KCl 溶解的累计时间；n_2 为所加溶质 KNO_3 或 KCl 的累计物质的量；$W_{累}$ 为所加溶质 KNO_3 或 KCl 的累计质量；M_2 为 KNO_3 或 KCl 的摩尔质量。

本实验用到的 WLS-2 恒流电源及 SWC-ⅡD 温度温差仪使用方法见第 4 章的 4.11、4.12。

三、仪器与试剂

WLS-2 数字恒流电源；SWC-ⅡD 数字温度温差仪；量热器（含加热器）；磁力搅拌器；KNO_3（AR）或 KCl（AR）。

四、实验步骤

1. 在台秤上称量 8 份 KNO_3 或 KCl，质量分别为 1.5g，2.0g，2.5g，3.0g，3.5g，4.0g，4.5g，5.0g，并编号 1-8。

2. 用量筒直接量取 216.2mL 蒸馏水加入量热器，按图 2.7 连好线路。

3. 检查无误后，调节电源（先将 WLS-2 恒流电源粗调、细调旋钮逆时针旋到底，再打开电源）。调节电流，使电流 I 和电压 U 的乘积（$P=IU$）约为 2.5W。

4. 打开温度温差仪电源和搅拌器电源。当水温上升到比室温水高出 0.5℃时，按采零键并按锁定键锁定，开始计时，从加料口加入第一份样品，然后塞好塞子。记录电压和电流。加入样品后温度会下降，并变为负温差。

5. 当温差值由负变为零时，记下此时加热时间 t_1，并加入第二份样品，重复上述步骤继续测定，直至 8 份样品全部加完为止。

6. 测定完毕后，切断电源，打开量热器，检查 KNO_3 或 KCl 是否溶完（如未溶完，则必须重做）。将量热器等洗净放回原处。

五、注意事项

1. 因加热器开始加热时有一滞后性，故应先让其加热正常，使温度高于环境温度 0.5℃左右，开始加入第一份样品并计时。

2. 实验过程中要求 $P=IU$ 值恒定，故应随时注意调节。

3. 实验过程中的计时为累计计时，故要注意连续计时。

4. 为确保 KNO_3 或 KCl 迅速、完全溶解，在实验前先将 KNO_3 或 KCl 颗粒研细。

5. 整个测量过程要尽可能保持绝热，减少热损失，盖严量热器盖子及加料孔塞。

六、数据处理

1. 根据溶剂的质量和加入溶质的质量，求算溶液的浓度，以 n_0 表示（以溶质 KNO_3 为例）：

$$n_0 = \frac{n_{H_2O}}{n_{KNO_3}} = \frac{216.2}{18.02} \div \frac{W_{累}}{101.1} = \frac{1213}{W_{累}}$$

2. 按式(2.13)计算各次溶解过程的热效应。

3. 按每次累积的浓度和累积的热量，求各浓度下溶液的 n_0 和 Q_s，并作 Q_s-n_0 图。

4. 从 Q_s-n_0 图中求出 $n_0=80$，100，200，300 和 400 处的微分溶解热和微分冲淡热。

5. 计算 n_0 从 80 →100，100→200，200→300，300→400 的积分冲淡热。

6. 数据记录见表 2.8。

表 2.8 KNO_3 溶解热的测定

序号	KNO_3 质量 W/g	KNO_3 累积质量 W_a/g	加热时间 t/s
1			
2			
3			
4			
5			
6			
7			
8			

七、思考题

1. 如果过程是放热的，则应如何进行实验？
2. 何谓积分溶解热和微分溶解热？我们测的是哪种？
3. 影响本实验结果的因素有哪些？
4. 温度和浓度对溶解热有无影响？
5. 图 2.10 是一种电热补偿线路，你还能设计其他的电热补偿线路吗？但不论采用何种线路所得之 IU 值总是有一定的近似，若要精确知道 I、U 值，则又如何测定？

Experiment 2.3 Determination of Heat of Solution

Objectives

1. To learn the operation of a calorimetric apparatus and the technique of electrothermic compensation method in heat effect measurement.

2. To determine the integral heat of solution of potassium nitrate and to calculate the differential heat of solution, integral heat of dilution and differential heat of dilution.

Theory/Principle

1. Heat of solution The quantitative study of the thermal effects which accompany the solution of a solute in a pure solvent or a solution has been systematized through the instruction of the concepts of the general heat of solution, the integral and differential heat of solution.

The general heat of solution Q at a particular concentration is the heat of reaction at a specified temperature and pressure when n_2 moles of solute is dissolved in n_1 moles pure solvent or a solution to procedure a solution of the given concentration. The integral heat of solution Q_s is the heat of reaction at a specified temperature and pressure when 1 mole of solute is dissolved in n_0 moles pure solvent or a solution. For a solution of given concentration, the differential heat of solution.

The solute is the heat of solution per mole of solute added under conditions in which the concentration of the solute is changed only differentially. Thus, if δq represents the enthalpy change when dn_2 moles of solute is added (at constant T,p) to a solution phase already containing n_2 moles of solute and n_1 moles of solvent, the differential heat of solution is

$$\left(\frac{\partial Q}{\partial n_2}\right)_{n_1}$$

The heat of dilution is the heat effect, at constant temperature and pressure, accompanying the addition of 1 mole solvent to a quantity of solution of a particular concentration to procedure a lower concentration solution.

The integral heat of dilution $Q_{d, n_{01} \to n_{02}}$ between two concentration n_{01} and n_{02} is defined as the heat effect, at constant temperature and pressure, accompanying the addition of enough solvent to a quantity of solution of concentration n_{01} containing *one mole of solute* to reduce the concentration with the higher value n_{02}. The *differential heat of dilution* of the solute the heat of solution per mole of solvent added into the solution of a certain concentration and limiting quantity.

In this experiment, we use the electrothermic compensation method to determine the thermal effect. The integral heat of solution can be determined directly, and the other three thermal effect will be calculated through the curve of Q_s (the integral heat of solution) against n_0 (the ratio of the mole number of the solvent to that of the solute). Plot the Q_s, n_0 curve, we may get a graph like Fig. 2.11.

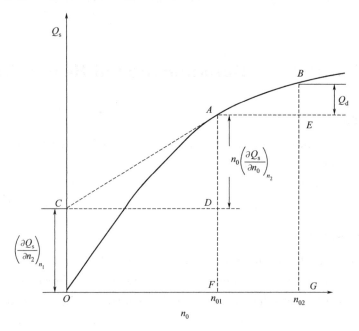

Fig. 2.11 Curve of integral heat of solution against n_0

In Fig. 2.11, the segment AF and BG represent the integral heat of solution accompanying *one mole of solute* dissolving into n_{01} and n_{02} *moles of solvent* apart. Thus the segment BE represents the integral heat of dilution for the process in which some solvent is added to a solution already containing the fixed number of 1 mole of solute and n_{01} moles of solvent, and then the mole numbers of the solvent increase from n_{01} moles to n_{02} moles with the result of dilution of the solution. So the integral heat of dilution of dissolution $Q_d = (Q_s)_{n_{02}} - (Q_s)_{n_{01}} = BG - EG$.

From the definition of the differential heat of dilution, we can see that the slope of the plot of the integral heat of solution at point A in Fig. 2.11 equals the value of the differential heat of dilution.

$$\left(\frac{\partial Q}{\partial n_1}\right)_{n_2} = \left(\frac{\partial Q/n_2}{\partial n_1/n_2}\right)_{n_2} = \left(\frac{\partial Q_s}{\partial n_0}\right)_{n_2} = \frac{AD}{CD} \tag{2.9}$$

Since the definition of the integral heat of solution requires $Q = n_2 Q_s$, the intercept of the slope on the vertical axis is the differential heat of solution at the specified composition.

$$\left(\frac{\partial Q_s}{\partial n_2}\right)_{n_1} = Q_s - n_0 \left(\frac{\partial Q_s}{\partial n_0}\right)_{n_2} = OC \tag{2.10}$$

2. Electrothermic compensation method

The features of an diabatic calorimeter used in this experiment are shown in Fig. 2.12. The

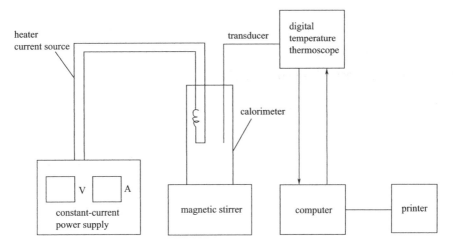

Fig. 2.12 Calorimetric apparatus for measuring heats of solution

process of the dissolution of potassium nitrate into water is an endothermic reaction. To compensate the temperature decrease, electrical energy is dissipated to raise the system temperature. Applying a current across the solution until the end of the dissolution reaction and the system temperature revert to initial value, the quantity of the heat required to compensate the temperature decrease can be calculated as follows

$$Q = I^2 Rt = UIt \tag{2.11}$$

Where Q is the heat dissipated or the heat effect of the solution in joules; I is the current in amperes; R is the resistance in ohms of the heating element; t is the time in seconds of current flow; U is the voltage drop in volts across the resistance wire.

Apparatus and Chemicals Required

Apparatus WLS-2 constant-current power supply with digital output, SWC-II$_D$ digital temperature thermoscope, calorimeter including electric heater and glass flask, magnetic stirrer, cylinder, timer, table balance.

Chemicals KNO_3 (AR).

Procedure

1. Weigh out eight samples of KNO_3 with the balance to about 1.5, 2.0, 2.5, 3.0, 3.5, 4.0, 4.5 and 5.0g and number them from No. 1 to No. 8 respectively.

2. Weigh out 216.2mL of water into the glass flask with cylinder. Assemble the calorimeter and connect the circuit as indicated in Fig. 2.12. Record the temperature of the water and take it as the experimental temperature.

3. Switch on the power supply, and set the electric power to about 2.5W.

4. Switch on the magnetic stirrer to a moderate rate. Record the voltage and current data. When the digital temperature thermoscope indicates a 0.5℃ increase of temperature, press the key of returning to zero and take it as starting temperature. And at the same time press the timer to record the time.

5. Add the first sample of KNO_3 as quickly as possible. Then the temperature difference will drop to a negative value. When it returns to zero from the negative value, record the heating time (remember not to stop the timer). Then add the second sample of KNO_3.

6. Repeat procedure 5 till all the eight samples of KNO$_3$ are added.

7. Switch off the power, and check the dissolution of KNO$_3$ (redo the experiment if KNO$_3$ is not dissolved completely). Clean the calorimeter and put it back.

Observations and Measurements

Table 2.9 Determination of Heat of Solution of KNO$_3$

Volume of water used ___ mL Room temperature ___ ℃
Current(I) ___ A Voltage(U) ___ V

Number of the samples	Mass of KNO$_3$ (W)/g	Accumulated mass of KNO$_3$ (W_a)/g	Heating time t/s
1			
2			
3			
4			
5			
6			
7			
8			

Calculations

1. According to the mass of the solute and the solvent, calculate the concentration of the solution denoted as n_0 as following

$$n_0 = \frac{n_{H_2O}}{n_{KNO_3}} = \frac{216.2}{18.02} \div \frac{W_a}{101.1} = \frac{1213}{W_a}$$

2. Calculate the heat effect of solution Q under each concentration according to equation 2.11.

3. Calculate the integral heat of solution Q_s under each concentration, where $Q_s = Q/n_2$, n_2: the accumulated moles of KNO$_3$.

4. Plot Q_s against n_0.

5. Calculate the differential heat of solution and dilution at $n_0 = 80$, 100, 200, 300 and 400.

6. Calculate the integral heat of dilution when n_0 changes from 80 to 100, from 100 to 200, from 200 to 300, and from 300 to 400.

Notes

1. Since the heater has hysteresis quality, we must preheat it till the 0.5℃ increase of temperature.

2. To keep the value of electric power ($P=UI$) invariable, we must adjust the voltage and the current to suitable values at any moment.

3. Remember to record the time continuously.

4. Pre-grinding KNO$_3$ is essential for the rapid and complete solution of solute.

5. To keep the system as adiabatic as possible and to prevent the heat exchange between

the system and the surroundings, we must stuff up the charging hole and the lid.

Questions/Exercises

1. How should we measure the heat of solution for an exothermic reaction?

2. What is integral heat of solution and differential heat of solution? And which is the one we measure in the experiment?

3. What factors will influence the result of the experiment?

4. Will the heat of solution depend on the temperature and the concentration?

5. Can you design another electrothermic compensation circuit diagram except that shown in Fig. 2.12? Whatever kind of circuit we use, there must be an approximation of the value of the voltage and the current. What should we do to record them precisely?

实验 2.4　燃烧热的测定

一、实验目的

用氧弹式量热计测量有机物的燃烧热；明确燃烧热的定义，了解 Q_V 与 Q_p 的差别；了解量热计中主要部件的作用，掌握量热计的使用技术；学会雷诺图解法。

二、实验原理

在适当的条件下，许多有机物都能迅速完全地进行氧化反应，这就为准确测定它们的燃烧热创造了有利条件。

氧弹是一特制的不锈钢容器，如图 2.13 所示。为保证样品在其中迅速而完全地燃烧，需要用过量的强氧化剂，通常氧弹中充以氧气作为氧化剂。实验时氧弹是固定在装有一定量水的不锈钢桶中，水桶外是空气隔热层，最外面是恒定的水夹套，如图 2.14 所示。

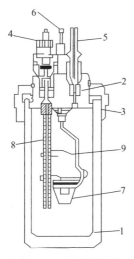

图 2.13　氧弹的构造

1—厚壁圆筒；2—弹盖；3—螺帽；4—进气孔；
5—排气孔；6—电极；7—燃烧嘴；
8—电极（同时也是进气管）；9—火焰遮板

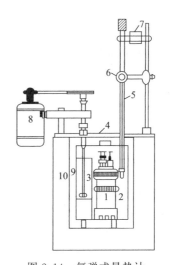

图 2.14　氧弹式量热计

1—氧弹；2—钢水桶；3—搅拌器；4—胶木盖；
5—贝克曼温度计；6—放大镜；7—振动器；
8—电动机；9—空气隔热层；10—水夹套

引火丝及样品在体积固定的氧弹中燃烧所放出的热大部分为水桶中的水吸收；其余部分为氧弹、水桶、搅拌器及感温探头等设备吸收。在量热计与环境没有热交换的情况下，可以写出如下热量平衡式：

$$-\frac{m}{M}Q_{V,m} = C_{if}\Delta T + Q_{丝}m_{丝} + Q_{N,m}n \tag{2.15}$$

式中，$Q_{V,m}$ 为样品的恒容摩尔燃烧热，$J \cdot mol^{-1}$；m 为样品的质量，kg；M 为样品的摩尔质量，$kg \cdot mol^{-1}$；C_{if} 为量热计热容，它包括氧弹、量热计及水的热容，$J \cdot K^{-1}$；ΔT 为准确温差，K；$Q_{丝}$ 为点火丝燃烧热，铁丝为 $-6696.4 \; kJ \cdot kg^{-1}$、镍丝为 $-3158.9 \; kJ \cdot kg^{-1}$、Cu-Ni 丝为 $-3136.2 \; kJ \cdot kg^{-1}$；$m_{丝}$ 为燃烧掉的点火丝质量（kg）即原丝质量减去燃烧剩余质量；$Q_{N,m}$ 为氮气被氧化成 1mol HNO_3 的反应热，$J \cdot mol^{-1}$；n 为生成 HNO_3 的物质的量，mol，用 $0.1 mol \cdot dm^{-1}$ 标准 NaOH 溶液滴定求得。

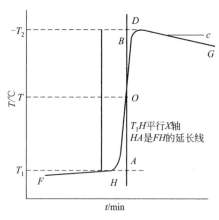

图 2.15 求真实温差的雷诺图解法

由式（2.15）知，要测量样品的 $Q_{V,m}$ 必须先知量热计的热容 C_{if}，测定的方法是用一定量已知燃烧热的标准物质（常用苯甲酸，$Q_{V,m} = -3230.6 \; kJ \cdot mol^{-1}$）在相同条件下进行试验，测量其温差，经校正为真实温差后代入式（2.15），算出 C_{if} 值。

关于真实温差的求算：

氧弹量热计不可能是严格绝热的。在燃烧后升温阶段，系统和环境间难免要发生热交换，因而温度计读得的温差并非真实温差。应作如下校正：通常样品燃烧后温升为 1.5～2.0℃，在燃烧前后观测水温变化，将水温对时间作图，连成折线 $FHOD$，如图 2.15 所示。图中 H 点相当于开始之点，D 点为观测到的最高温度。对 H 点对应的温度 T_1 和 D 点对应的温度 T_2 的平均 $\frac{T_1+T_2}{2}$ 为 T，经 T 点作横坐标的平行线 TO 与折线交于 O，然后过 O 点作垂直线分别与 FH 和 GD 交 A、B 两点，这两个交点所示间隔温度即所求真实温差 ΔT。

三、仪器与试剂

氧弹量热计（附压片机一台）；分析天平 1 台；1L 容量瓶一个；氧气瓶、氧气表各一个；50mL 滴定管 1 支；150mL 锥形瓶 1 个；分析纯苯甲酸；分析纯萘；引火丝 10cm；$0.0100 mol \cdot L^{-1}$ NaOH 标准溶液；酚酞指示剂。

四、实验步骤

1. 量热计热容 C_{if} 测定

（1）样品压片　剪取约 10cm 长引火丝在分析天平上称量，并精确测量其实际长度。将约 1g 苯甲酸（不可超过 1.1g），用压片机压成片，再称准至 0.0002g 放入坩埚内。

（2）装氧弹　拧开弹盖，将弹盖放在专用架上。把盛有苯甲酸的坩埚固定在坩埚架上，将点火线弯成 U 形，U 形的底部与片状苯甲酸紧密接触，再将点火线的两端固定在电极上。点火丝勿接触坩埚。氧弹内装 10mL 蒸馏水，拧紧氧弹盖子。用充氧机在 2.0MPa 的压力下充氧。

(3) 燃烧与测量　用容量瓶量取 3000mL 自来水装入量热计内桶，将氧弹装入量热计内桶。插好电极，盖上量热计盖子，插入测温探头。测温探头和搅拌器均不得接触氧弹和内桶。打开控制箱搅拌，设定信号响为 1min 一次。待 5min 系统稳定后开始测量温度。每隔 1min 读取温度一次，共读取 8 次，此为测量前期；按下点火键点火，然后开始读取主期温度，每 1min 读一次直到温度不再上升而开始下降为止；每隔 1min 读一次温度，共读 8 次，此为测量后期。

(4) 实验停止后，取出氧弹。用放气帽缓缓压下放气阀，放尽氧弹内气体，拧开并取下氧弹盖，检查燃烧情况（如果弹内有炭黑或未燃尽的试样微粒，实验失败）。取出剩余点火丝称重。倒出量热计中的水，洗净氧弹，用纱布把各部件擦干。称量剩下的点火丝重量。

2. 待测物（萘）的燃烧热测定

萘的用量约 0.7g（不可超过 0.8g），方法同前。

五、数据处理

1. 实验数据记录见表 2.10。绘出温度与时间曲线，用作图法求真实温差。由式 (2.15) 算出量热计热容 C_{if}。
2. 用同样的方法求萘燃烧的真实温差，并计算萘的恒容摩尔燃烧热 $Q_{V,m}$。
3. 计算萘的恒压摩尔燃烧热 $Q_{p,m}$。

表 2.10　实验数据记录

m(苯甲酸)：　　　　　　　　　　　　　　m(萘)：
m(燃烧前金属丝)：　　　　　　　　　　　m(燃烧后金属丝)：
L(燃烧前金属丝)：
室温：　　　　　　　　　　　　大气压：

测 C_{if} 时温度读数		测萘 Q_V 时温度读数	
序号	温度/℃	序号	温度/℃
前期 1		前期 1	
...		...	
8		8	
主期 1		主期 1	
2		2	
...		...	
后期 1		后期 1	
...		...	
8		8	

六、思考题

① 写出恒容摩尔燃烧热 $Q_{V,m}$ 与恒压摩尔燃烧热 $Q_{p,m}$ 的关系式。
② 用电解水制氧进行本实验，可以吗？
③ 本实验采用哪些绝热措施？

Experiment 2.4　Determination of Heat of Combustion

Objectives

Measuring the heat of organics combustion by the oxygen-bomb calorimeter; clearing the definition of heat of combustion; understanding the difference between $Q_{V,m}$ and $Q_{p,m}$, and the role of the main components of calorimeter; mastering the use of calorimeter, and Renault graphic institute.

Theory/Principle

Many organic compounds can quickly be oxidated under suitable conditions, thus their heat of combustion may be accurately measured

Oxygen bomb is a specially designed stainless steel container, as shown in Fig. 2.16. To ensure the sample in oxygen bomb to burn quickly and completely, the excessive strong oxidant is necessary. Usually oxygen is used as the oxidant. Oxygen bomb is equipped in the stainless steel barrel with a certain amount of water, and the outside is the insulating layer of air. Then the outside is a fixed water jacket, as shown in Fig. 2.17.

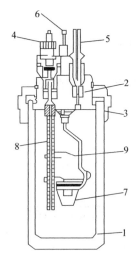
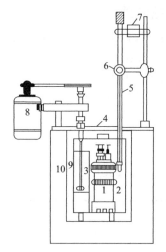

Fig. 2.16 oxygen bomb construction
1—thick-walled cylinder; 2—shell covered;
3—nut; 4—air intake; 5—exhaust vent;
6—electrode; 7—combustion crucible;
8—electrode; 9—flame cover plate

Fig. 2.17 oxygenbomb calorimeter
1—oxygen bomb; 2— bucket; 3—agitator;
4—bakelite covered; 5—Beckman thermometer;
6—magnifying glasses; 7—vibrator; 8—motor;
9—air insulation layer; 10—water jacket

The majority of the heat released by combustion of ignition wire silk and sample in the oxygen bomb with an fixed volume are absorbed by the water in the bucket, while the rest are absorbed by the equipment such as the oxygen bomb, the bucket, the blender and the temperature probe. Ignoring the exchange of heat between the calorimeter and the environment, the heat balance formula is as follows:

$$-\frac{m}{M}Q_{V,m} = C_{if}\Delta T + Q_{wire}m_{wire} + Q_{N,m}n \quad (2.15)$$

Where $Q_{V,m}$ is the constant-volume molar heat of combustion, J·mol^{-1}; m is the mass of the sample, g; M is sample molecular weight, g·mol^{-1}; C_{if} is calorimeter heat capacity, which includes heat capacity of the oxygen bomb, the calorimeter and the water, J·K^{-1}; ΔT is ideal temperature change, K; Q_{wire} is the heat of combustion of ignition wire, iron wire for -6696.4 J·g^{-1}, nickel wire for -3158.9 J·g^{-1}, Cu-Ni wire for -3136.2 J·g^{-1}; m_{wire} is the wire mass burning off (g), namely the mass of the original wire subtracting the residual mass; $Q_{N,m}$ is the reaction heat that nitrogen is oxidated into 1 mol HNO_3, $J \cdot mol^{-1}$; n is the molar quantity of HNO_3 formation, mol.

Knowing from formula (2.15), C_{if}, the calorimeter heat capacity, must be firstly obtaineded before the $Q_{V,m}$ of the sample is measured. The experiment is carried under the same conditions with a standard material the heat of combustion of which is known (commonly using benzoic acid, $Q_{V,m} = -3230.6$ kJ $\cdot$ mol^{-1}), the calorimeter heat capacity C_{if} may be figured out by calculating the ideal temperature change ΔT.

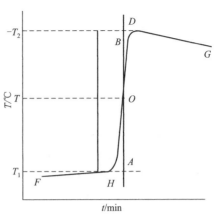

Fig. 2.18 Renault graphical method

The Calculation of the ideal temperature change ΔT:

Oxygen bomb calorimeter is not strictly adiabatic. It is inevitable that there is the interthermal exchange between the system and the environment. After combustion, thus, the temperature change is not ideal. It should be corrected as follows: as shown in Fig. 2.18. H-point means the beginning temperature of combustion, D-point means the maximum temperature observed. The temperature of H-point is T_1 and the temperature of D-point is T_2; T is the average value of T_1 and T_2. Drawing the parallel line through T-point to the abscissa, the intersection with the curve is in O. Drawing the parallel line through O-point to the longitudinal coordinate, the intersection with the extension line of FH and GD is in A and B Respectively.

$$\Delta T = T_B - T_A$$

Apparatus and Chemicals Required

An oxygen bomb calorimeter; a compression machine; an analytical balance; an oxygen bottle; a manometer of oxygen; pure benzoic acid (A.R.); pure naphthalene (A.R.); wire silk about 10 cm.

Procedure

1. Measurement of calorimeter heat capacity C_{if}

(1) Compression of sample: weighing ignition wire silk about 10 cm long in the analytical balance, and measuring its length accurately. Pressing about 1 g benzoic acid (not more than 1.1 g) into tablet with the machine, then weighing it accurately within 0.0002 g error, and putting it in the crucible.

(2) Installation of oxygen bomb: screwing the cover shell of oxygen bomb open, and putting it on the dedicated shelve. Placing the crucible containing benzoic acid on the crucible shelf. Bending ignition wire silk into U-shape, so that it is closely contacted with benzoic acid. Then fixing the two ends of ignition wire silk on the two electrodes respectively. Ignition wire can not be contacted with the crucible. Putting 10 mL of distilled water into the oxygen bomb, then tightening the oxygen bomb lid. Using oxygenation machine to supply oxygen under the oxygenation pressure of 2.0 MPa.

(3) Combustion and measurement: Putting 3000 mL of water into the barrel of the calorimeter; placing the oxygen bomb calorimeter into water in the barrel. Inserting electrodes properly, covering the calorimeter with the lid, and inserting the temperature probe. The temperature probe and the agitator can not be contacted with the oxygen bomb and the barrel. Setting the control box to ring once in a minute. The measurement will be carried when the system temperature is nearly stable. Reading the temperature every one minute, eight times' reading in total, and this is the early-period of the measurement; then pressing the ignition button, and reading the temperature every one minute until the temperature no longer to rise, and this is the main-period of the measurement; when the temperature begin to decline, reading the temperature every one minute also, eight times' reading in total. This is the late-period of the measurement.

(4) Removing the oxygen bomb out when the experimental is over, and releasing oxygen with deflated cap, then screwong the oxygen bomb cap open, observing if the sample has been burned completely (if there were carbon black or the unburned sample, the experiment had failed). Weighing the remained ignition wire, disposing the water in the calorimeter, cleaning the oxygen bomb, and rubbing the components dry by gauze. weighing the residual wiresilk.

2. Determination of the heat of combustion of naphthalene

Weighing naphthalene about 0.7 g (not more than 0.8g). Carrying on the experiment with the same method instead of benzoic acid.

Calculations

1. Drawing the temperature and the time curve, and caculating the ideal temperature change with Renault graphical method (Table 2.11). Figuring out the calorimeter heat capacity C_{if} by formula (2.15).

Table 2.11 Observations and measurements

m (benzoic acid): L (wire silk before combustion):
m (naphthalene): m (wire silk before combustion):
m (wire silk after combustion):
room temperature: atmospheric pressure:

benzoic acid		naphthalene	
No.	$t/℃$	early 1	$t/℃$
early 1		...	
...		8	
8		main 1	
main 1		2	
2		...	
...		late 1	
late 1		...	
...		8	
8		early 1	

2. Using the same method to measure the ideal temperature change for combustion of naphthalene, and figuring out $Q_{V,m}$, the constant-volume molar heat of combustion of naphthalene.

3. Figuring out $Q_{p,m}$, the constant-pressure molar heat of combustion of naphthalene.

Questions/Exercises

1. Writing the relationship between $Q_{p,m}$ and $Q_{V,m}$.

2. Whether this experiment can be carried on with oxygen obtained by water electrolysis?

3. What heat insulation measurements have been made in this experiment?

实验 2.5 液体饱和蒸气压的测定

一、实验目的

用等压计法测定乙醇的饱和蒸气压,图解法求乙醇的平均摩尔蒸发焓和正常沸点。理解饱和蒸气压与温度的关系,掌握等压计法的原理。验证 Clausius-Clapeyron 方程。

二、实验原理

纯液体的饱和蒸气压随温度的升高而增大,其定量关系可用 Clausius-Clapeyron 方程表示:

$$\frac{\mathrm{d}\ln p}{\mathrm{d}T} = \frac{\Delta H}{RT^2} \tag{2.16}$$

式中,p 为纯液体的饱和蒸气压;T 为热力学温度,K;ΔH 为液体的摩尔蒸发焓,J·mol^{-1};R 为气体通用常数,8.314J·mol^{-1}·K^{-1}。蒸发焓一般随温度的上升而降低,当温度范围变化不大时 ΔH 可视为常数,积分式(2.16)得:

$$\ln p = -\frac{\Delta H}{RT} + C \tag{2.17}$$

式中,C 为积分常数,实验测定液体在不同温度下的饱和蒸气压,作 $\ln p$-$\frac{1}{T}$ 图为直线,由直线斜率可求出 ΔH。

测定液体饱和蒸气压常用的方法有两类:

(1) 静态法　适用于蒸气压较大的液体;

(2) 动态法　一般适用于蒸气压较小的液体。

本实验依据静态法,采用等压计测定不同外压下乙醇的沸点,其原理如下。

等压计(见图 2.19)系由相互连通的三个玻璃管组成。在管 a 内装有待测液体,管 b 和管 c 中的液体(待测液)将管 a 与外部空气隔开,管 a 中液体蒸气所产生的压力 p^*(即饱和蒸气压)作用于管 c 液面上。当管 b 与管 c 液面平齐时,表示蒸气压 p^* 与管 b 上部外压 $p_{外}$ 相等,记下此时的外压和温度,即为待测液体在该温度下的饱和蒸气压(外压可由压差计调节)。

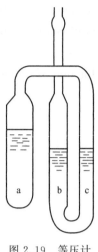

图 2.19 等压计

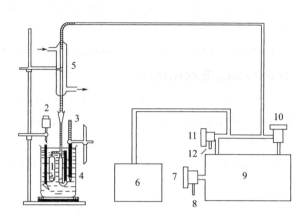

图 2.20 饱和蒸气压测定装置图

1—等压计；2—搅拌器；3—水银温度计；4—水浴；5—冷凝器；
6—精密压差计；7—抽气阀；8—接真空泵；9—压力缓冲罐；
10—平衡阀；11—阀；12—大气

$$p^* = p_{外} = p_{大气} - \Delta p$$

式中，Δp 为压差计上的压力差。

三、实验装置

饱和蒸气压测定装置全套；抽气泵 1 台；无水乙醇（AR）。

图 2.20 为液体饱和蒸气压测定的装置。图 2.19 中管 b 上方外压接真空泵抽气并由精密压差计控制，其间连接不锈钢缓冲罐起稳压作用；干燥塔和缓冲瓶起保护真空泵的作用；等压计置于大烧杯水浴中（应全部浸于水浴中），由插入其中的温度计读取温度；电炉用于加热水浴；冷凝管将蒸气冷凝回流至等压计中。

该装置的优点在于：①对等压计作了改进，使平衡易于观察，误差小；②用不锈钢缓冲罐代替以往使用的玻璃缓冲瓶，其中通大气的阀门 11，抽气阀 7，并在两者之间增加了一个平衡阀 10。在实验操作减压前，先对缓冲罐减压（储存真空），实际减压时，开平衡阀 10 即可对系统减压，操作简便且避免真空泵长时间开启产生噪声和浪费能源。

四、实验步骤

1. 实验准备

将等压计（图 2.19）洗净烘干，装入适量的乙醇，其中管 a 约 2/3，管 b、c 中约 1/2，将等压计按图 2.20 连接好。

2. 检漏与真空储存

接通精密压差计（参见 "4.5 精密数字压力计"）电源，打开压力缓冲罐上所有阀门（即系统与大气相通），开真空泵，关闭通大气的阀门 11，此时系统开始减压。当精密压差计显示压差为 50~60 kPa 时，关闭抽气阀 7，静观精密压差计之数据有无明显变化以确定系统是否漏气。如不漏气，应立即关闭平衡阀 10（使真空储存在缓冲罐内），打开通大气的阀门 11（使系统置于大气压下）。饱和蒸气压减压装置的操作参见 "4.3 饱和蒸气压减压装置"。

3. 驱逐空气

经检查不漏气后，旋转阀 11 使系统通大气，打开冷凝水，电炉通过电加热并搅拌水浴，使等压计内液体升温至 76℃（关电炉）并沸腾 3min，即可逐出管 a、c 上部空间中的空气，但切勿使水温升得过高，否则蒸气来不及冷凝腐蚀橡皮管造成系统漏气。由于加热时间较长，此步骤可与步骤 2 同时进行，但应注意真空储存完后必须及时打开通大气的阀门 11，以免系统在高真空下加热使待测液体蒸发损失过多。

4. 降温测定

当停止加热时，管 b 的液面将高于管 c 的液面（为什么?），搅拌水浴使水浴温度均匀下降，蒸气压 p^* 随温度下降而减少，管 c 液面上升，管 b 液面下降，当两液面相平齐时（$p^* = p_{外} = p_{大气}$，此时 $\Delta p = 0$），立即读出水浴温度（事先应一直观察温度计读数的变化，否则来不及读数），同时关闭通大气的阀门 11（以后均不再打开!），缓开平衡阀 10 使压差计上减压 3~4kPa 后立即关闭（注意减压必须及时，否则因 p^* 下降较快使 $p_{外} - p^*$ 大于管 b、c 中的液柱差时，空气将倒灌于 a、c 上部空间，导致必须重复驱气步骤）。及时抽气减压是确保实验成功的关键。

搅拌水浴，当管 b、c 两液面平齐时记下水浴温度和压差计之压力差，立即进行下一次减压。每次递减压力 3~4kPa，重复上述步骤共测 8~10 组数据。

注意实验前后记录大气压力（气压计的使用参见"4.1 大气压力计"）。

五、注意事项

1. 当使用图 2.20 的测定装置时，应清楚各阀门的功能用途，操作阀门时动作应轻、缓，否则易使阀门内橡胶密封圈损坏而导致漏气，且每次阀门应尽可能置于合适位置以便于下次操作。

2. 不宜在加热水浴时检查漏气，因加热时蒸气压变化使压力差 Δp 不稳定，本实验方法中如系统有很小漏气对结果影响不大（为什么?）。

3. 降温测定时应不停地搅拌使水浴温度均匀，不宜添加冷水使降温速度过快，以免出现假平衡状态使结果误差增大。

4. 本实验之关键在于两管液面平齐时读取数据，且必须使观察等压计两管液面、搅拌水浴、读取温度和压差计上的压力差（Δp）及抽气减压五项动作同时进行（每次抽气减压均为下一组数据的测定作准备）。因此实验前同组的同学一定要分工负责，相互配合。当管 b 液面不冒气泡时即将接近平衡。

5. 若压差计有起始误差 Δh 时，实际压力差 $\Delta p = \Delta h_{观} \pm |\Delta h|$。取"+"、取"−"视具体情况而定。

表 2.12 液体饱和蒸气压的测定实验数据记录

大气压：　始　　　　末　　　　平均值　　　　室温：

编号	温度 ℃	温度 K	Δp/kPa	$\dfrac{1}{T} \times 10^3$	饱和蒸气压 p /kPa	$\ln p$
1						
2						
3						
4						
5						
6						
7						
8						
9						
10						

六、数据处理

1. 数据记录见表 2.12。作蒸气压-温度图,由图中曲线求出乙醇的正常沸点 T_b。

2. 作 $\ln p - \frac{1}{T}$ 图,由直线斜率计算在实验温度范围内乙醇的平均摩尔汽化热 ΔH,并与文献值($\Delta H = 40.3 \text{kJ} \cdot \text{mol}^{-1}$)进行比较。

要求同时用计算机处理和手工作图处理,并比较之。

七、思考题

1. 分析误差及其产生的原因。
2. 实验测定之前为何要驱逐空气?
3. 如果未及时抽气减压而造成空气倒灌,将对实验结果有何影响?
4. 等压计之水浴温度下降过快对结果有何影响?

Experiment 2.5　Determination of Saturated Vapor Pressure of a Pure Liquid

Objectives

1. To determine the satarated vapor pressure for liquid of alcohol with the isoteniscope.
2. Determination of the molar enthalpy of vaporization and the normal boiling opoint of alcohol. To demonstrate the validity of Clausius- Clapeyron equation.

Theory

Vapor pressure varies exponentially with the reciprocal of temperature. The Clausius-Clapeyron equation expresses the relation between vapor pressure, temperature, and the molar enthalpy of vaporization.

$$\frac{d\ln p}{dT} = \frac{\Delta H}{RT^2} \tag{2.16}$$

and

$$\ln p = -\frac{\Delta H}{RT} + C \tag{2.17}$$

Where p is vapor pressure; T is Kelvin temperature; ΔH is molar enthalpy of vaporization; R is gas constant.

The simplest method of measuring the vapor pressure is to connect a manometer to the liquid-vapor system. This is the static or direct method. Foreign gases and volatile impurities are removed by boiling the liquid.

A simple device for measuring the vapor pressure of liquid is the isoteniscope (Fig. 2.21). The liquid was injected to tube (a), the liquid of inside U-tube (b and c) separated from the external air from tube (a).

When the manometric liquid is at the same level in both limbs of the internal U-tube, the pressure observed with an external manometer is the saturated vapor pressure.

Apparatus and Chemicals Required

A device of measurement vapor pressure (Fig. 2.22); a vacuum pump; anhydrous alcohol (A. R.).

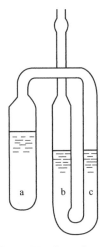

Fig. 2.21 Isoteniscope

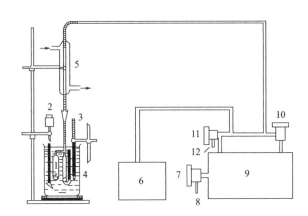

Fig. 2.22 Device of measurement vapor pressure
1—isoteniscope; 2—stirring apparatus; 3—thermometer; 4—beaker;
5—condenser; 6—digital pressure gauges; 7—vacuum valve;
8—to vacuum pump; 9—pressure buffer tank; 10—equlibrium valve;
11—atomsphere valve; 12—to atmosphere

Procedure

Clean and dry the isoteniscope, and assemble as shown in Fig. 2.22. Fill tube (a) about two-thirds full of the alcohol and half of (b) and (c).

Apply suction to the apparatus, close (12), and see that there are no leaks. If the apparatus is tight, admit air through (12). Turn on the atomsphere valve (12) so that the system is open to the atmosphere, commence stirring, and heating the bath until the liquid in the tube commences to boil, continue the boiling long enough to expel all air from the space between the main body of the liquid and the trap, then carry the temperature up to the highest value at which it is desired to determine the vapour pressure (78℃).

By manipulating valves (11) and (12), admitting or removing air as needed, the liquid levels in the two arms of the trap (b and c) are kept at the same height. It is very important that air is not permitted to be sucked back into the bulb.

When the heat is turned off, and as the system cools down, the liquid in the trap is kept level by manipulation of the two valves. Simultaneous readings of the manometer and the thermometer are taken, and supply the necessary data for the vapor pressure curve. These readings may be taken for once when the pressure is decreased 3~4 kPa each time. It should be measure 8 to 10 times totally.

Observations and Measurements

Table 2.13 barometer at start: barometer at end: room temperature:

No.	temperature		$\frac{1}{T} \times 10^3$	Δp/mmHg	vapor pressure p/mmHg	$\ln p$
	℃	K				
1						
2						
3						
4						
5						
6						
7						
8						
9						
10						

Calculations

1. Plot the vapor pressure-temperature curve; determination of the normal boiling opoint of alcohol from the curve (Table 2.13).

2. Plot the $\ln p - \frac{1}{T}$ curve; determination the molar enthalpy of vaporization of alcohol; using the equation of Clausius-Clapeyron.

Notes

1. Do not heat water bath before the check leakage.
2. If air gets back into the boiler between readings, it is necessary to boil the liquid again.
3. Stiring should be kept in the water bath for temperature uniformity.
4. The five movements (observe the pressure of two liquid, stirring bath, read the temperature and pressure of the Δp and decompression) must put up at the same time.

Exercises

1. Why it is necessary that expel all air from the space between the main body of the liquid and the trap?
2. Analysis the errors and the reasons.

实验2.6 二组分汽-液平衡相图的测定

一、实验目的

掌握二组分沸点-组成相图的测绘方法；掌握阿贝（Abbe）折光仪及超级恒温槽的使用方法。

二、实验原理

两种液态物质以任何比例混合都形成均相溶液的系统为完全互溶双液系。在恒定压力下溶液沸点与平衡的气、液相组成的关系，可用沸点-组成图（t-x图）表示。完全互溶双液系的沸点-组成图可分为三种：一种为最简单的情况，溶液沸点介于两个纯组分沸点之间，如图2.23所示。纵坐标表示温度，横坐标表示组分B的摩尔分数（x_B，y_B）。下面一条曲线表示汽-液平衡时温度（即溶液沸点）与液相组成的关系，称液相线（t-x线）。上面的线表示平衡温度与气相组成的关系，称气相线（t-y线）。若总组成为z_B的系统在压力p及温度t时达到汽-液两相平衡，其液相组成为x_B，气相组成为y_B（见图2.23）。另两种类型为具有恒沸点的完全互溶双液系统汽-液平衡相图，如图2.24所示。其中（a）为具有低恒沸点相图，（b）为具有高恒沸点相图。这两类相图中气相线与液相线在某处相切。相切点对应的温度称为恒沸点，对应组成的混合物称恒沸混

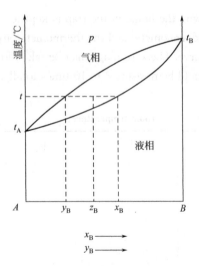

图2.23 正常的沸点-组成相图

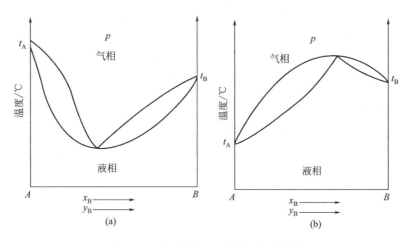

图 2.24 具有恒沸点的沸点-组成相图

合物。恒沸混合物在恒沸点达汽-液平衡，平衡的气、液组成相同。同一双液系在不同压力下，恒沸点及恒沸混合物是不同的。

本实验用气液沸点仪（见图 2.25）在一定压力下（通常在大气压力下），测定不同总组成（即加入平衡沸点仪溶液的组成）的环己烷和乙醇构成的溶液达到汽-液平衡时的温度及气、液相组成。根据这些数据作出该系统在此压力下的沸点-组成图。相图与压力有关，制作相图时必须注明其平衡压力值。

两种纯液体构成理想混合物时，其中各组分的汽-液平衡分压在所有浓度范围内都符合拉乌尔定律：

$$p_1 = p_1^* x_1 \qquad p_2 = p_2^* x_2 \qquad (2.18)$$

式中，p_1、p_2 为两组分汽-液平衡时气相分压；x_1、x_2 为平衡时两组分的液相物质的摩尔分数；p_1^*、p_2^* 为两组分纯液体在平衡温度下的饱和蒸气压。

若构成非理想混合物，其性质则不符合拉乌尔定律，在低压下可符合下列各式所示关系：

$$p_1 = p_1^* a_1 \qquad p_2 = p_2^* a_2 \qquad (2.19)$$
$$a_1 = \gamma_1 x_1 \qquad a_2 = \gamma_2 x_2 \qquad (2.20)$$

式中，a_1、a_2 为两组分在平衡液相中的活度；γ_1、γ_2 为两组分在平衡液相中的活度系数。

活度系数除与两组分本性有关外，还与温度、压力及组成有关。根据分压定律及式（2.19）、式（2.20）可得

$$\gamma_1 = \frac{p y_1}{p_1^* x_1} \qquad \gamma_2 = \frac{p y_2}{p_2^* x_2} \qquad (2.21)$$

式中，p 为汽-液平衡总压；y_1、y_2 为两组分在平衡气相中的摩尔分数。当从手册上查到一定温度下的 p_1^*、p_2^*，并从相图上查到平衡总压 p 及在此温度下汽-液平衡时的 x_1、y_1（x_2、y_2），就可用式（2.21）计算出各组分在平衡液相中的活度系数 γ_1、γ_2。

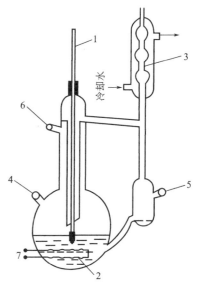

图 2.25 气液沸点仪

1—温度计；2—电热丝；3—冷凝管；
4—液相取样口；5—气相冷凝液取样口；
6—空气排出口；7—变压器接头

溶液的折射率随它的组成而改变。如果在一定温度下配制一系列已知组分的二组分溶液，测定出它们的折射率，可绘制出反映此溶液组成与折射率关系的标准曲线。根据此标准曲线可由此二组分溶液在给定温度下的折射率查得其组成。亦可由实验数据回归为组成与折射率的关系式，由关系式计算混合溶液在不同折射率下的组成。

三、仪器与试剂

沸点仪 1 套；超级恒温槽一台；阿贝折光仪一台；变压器一台；0.1℃刻度温度计(50～100℃) 1 支；30mL 小滴瓶 6 个；20mL 量筒 2 个；毛细滴管 2 支；洗耳球 1 个；擦镜纸若干。无水乙醇（AR），环己烷（AR），Ⅰ～Ⅷ号不同组成乙醇-环己烷溶液，丙酮。

四、实验步骤

1. 接好超级恒温槽与折射仪间的循环水管，将超级恒温槽温度调至 25.0℃。
2. 校正折射仪（使用和校正方法参见"4.4 阿贝折光仪"）。
3. 绘制标准曲线；用吸量管配制 6 种不同体积分数（10％、25％、40％、55％、70％、85％环己烷）各 10mL 的乙醇-环己烷溶液，分别放在 6 个干燥的 30mL 滴瓶中（注意盖严）。记录配制时室温。用折光仪分别测定各溶液的折射率（三学时实验不做此步，标准曲线数据由教师给出）。
4. 读取大气压力（参见"4.1 大气压力计"）。
5. 在干燥的沸点仪内，加入实验室配制好的乙醇-环己烷Ⅰ号溶液。使沸点仪内液面达到温度计水银球约一半的位置。开冷却水。沸点仪电热丝接至变压器 20V 的输出位置，加热至沸腾，使气相冷凝液充分回流。此时应注意观察温度。当在 2～3 min 内温度不变时，认为汽-液相达到平衡，记下温度数值。停止加热。迅速用干燥的毛细滴管先取气相冷凝液样品，测其折射率。用丙酮洗净折光仪棱镜后，再用另一支干燥的毛细滴管取液相样品，测定其折射率。洗净棱镜做好下次测定的准备工作。测定完毕，放出Ⅰ号溶液装入原溶液瓶内（切勿装错）。
6. 在沸点仪内加入Ⅱ号溶液，按上述步骤重复操作。再依次测量Ⅲ～Ⅷ号溶液。
7. 实验完毕，切断超级恒温槽电源，擦净折光仪。重读大气压。

五、数据处理

1. 在标准曲线上根据不同标号溶液的折射率查出相应的气、液相组成，如所测温度与标准曲线不符，则按温度每升高 1℃ 折射率下降 0.0005 校正后再查。
2. 用实测温度和汽-液平衡组成绘制乙醇-环己烷的 t-x 图，从图上查出最低恒沸组成和恒沸温度。
3. 计算含乙醇摩尔分数为 0.10 及 0.80 的乙醇-环己烷溶液在它们各自汽-液平衡温度下两组分的活度系数（两组分在平衡温度下的纯物质饱和蒸气压由附表 6 查出）。

六、注意事项

1. 变压器输出电压应缓缓上调至液体沸腾为止，一般不得超过 20V，否则会烧断电热丝。
2. 沸点仪中所加溶液必须使液面超过其内的电热丝玻管。
3. 每次测定前，取样管和折光仪棱镜玻璃面必须用洗耳球吹干，否则将严重影响测定结果。

七、思考题

1. 本实验中沸点仪及毛细滴管为什么必须干燥？本实验测得沸点-组成图的误差主要来源于哪些操作？

2. 在沸点仪内的系统中，为什么说总组成就是原始溶液组成？在达到汽-液平衡时，哪部分数量为平衡的气相量？哪部分为平衡的液相量？

3. 此实验中某一号溶液组成发生不大的变化，对实验测得相图是否有影响？

Experiment 2.6 Determination of Phase Diagram of a Two-component Liquid-vapor Equilibrium System

Objectives

1. To draw the phase diagram of cyclohexane-ethanol binary liquid-vapor system and to learn the basic concept of phase diagram and the phase rule.

2. To learn the method determining the composition of binary liquid system by Abbe refractometer.

Principle

A two-component liquid system is composed of a mixture of two liquids. If the two components are miscible in all proportions, it is called a totally miscible liquid. At a given pressure, the solution boiling point with the balance of the liquid-vapor composition relationship can be shown in a phase diagram of temperature vs. composition (t-x diagram). The boiling point - composition diagram of a totally miscible two liquids system can be divided into two or three types: the simplest type is the solution boiling point lies between the two pure component boiling point as shown in Fig 2.26.

The vertical axis indicates the temperature and the horizontal axis indicates the mole fraction of the component B (x_B, y_B). The below curve indicates the relationship of the liquid-vapor equilibrium temperature and liquid composition is liquid line; the above curve indicates the equilibrium temperature and vapor composition relations is the gas line. The liquid-vapor equilibrium phase diagram of totally miscible two-fluid system with a constant boiling point are the other two types as shown in Figure 2.27. Where (a) is that with a low constant boiling point of the phase diagram and (b) is that with a high constant boiling point of the phase diagram. These two types of phase diagram of the gas line and liquid line are tangent at somewhere. Tangent point corresponding to the temperature is called the temperature of the constant boiling, and the corresponding mixture is called constant boiling mixture. The constant boiling mixture reach the liquid-vapor equilibrium at the constant boiling point, and gas and

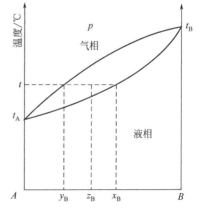

Fig. 2.26 Normal phase diagram of boiling point against the composition

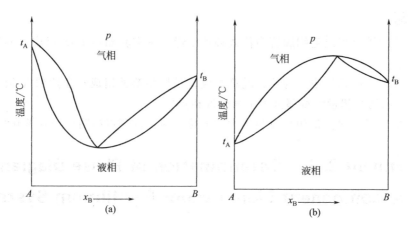

Fig. 2.27 The phase diagram of boiling point against the composition which has the constant boiling point

liquid have the same composition. If the pressure is changed, it would make some difference to the phase diagram.

In general, a liquid-vapor diagram can be drawn after measuring the boiling points and the compositions of the liquid and vapor phases of a series of binary mixtures.

When the ideal mixture is made up of the two pure liquids, which the gas-liquid equilibrium partial pressure of each component is in line with Raoult's law in all concentration range:

$$p_1 = p_1^* x_1 \qquad p_2 = p_2^* x_2 \qquad (2.18)$$

Where p_1、p_2 are gas partial pressure of two components (liquid-vapor equilibrium); x_1、x_2 are mole fraction of two components of liquid substance (liquid-vapor equilibrium).

If the mixture is not a ideal mixture, then its nature is not in accordance with Raoult's law, but it can comply with the following kinds of equation under low pressure.

$$p_1 = p_1^* a_1 \qquad p_2 = p_2^* a_2 \qquad (2.19)$$

$$a_1 = \gamma_1 x_1 \qquad a_2 = \gamma_2 x_2 \qquad (2.20)$$

Where a_1, a_2 are the activities of two components and γ_1, γ_2 are the actirity coefficients respectively.

Activity coefficient is not only related to the nature of the two components, but also it is relevant with temperature, pressure and composition.

$$\gamma_1 = \frac{p y_1}{p_1^* x_1} \qquad \gamma_2 = \frac{p y_2}{p_2^* x_2} \qquad (2.21)$$

The refractive index of the solution change with its composition. If measured the refractive index of a series of known components of two components solution at a certain temperature, we can draw the standard curve that reflects the composition of this solution and the refractive index. And we can also regress the relationship between composition and refractive index with the experimental data, and calculate the composition of the mixed solution under a different refractive index by the equation.

Apparatus and Reagent

Ebulliometer 1; ultra thermostatic water-circulator bath, 1; Abbe refractometer, 1;

voltage transformer, 1; thermometer, 1; 30 mL drop bottle, 6; 20mL graduated cylinder, 2; capillary burette, 2; rubber suction bulb, 1; lens wiping paper; absolute ethanol (AR); cyclohexane (AR); acetone (AR).

Procedure

1. Install the clean and dry ebulliometer according to Fig. 2.28. Set the ultra thermostatic water-circulator bath temperature to 25.0℃.

2. Correction of refractometer.

3. Making the standard curve (the standard curve is given by teachers).

4. Read atmospheric pressure.

5. Add the Ⅰ solution of ethanol-cyclohexane into a dry ebulliometer and add the liquid solution to one half of the position of the thermometer mercury ball. Open cooling water. Set the transformer to the 20V voltage and heat the liquid to boiling and make the gas phase condensate full reflux. Stop heating. Firstly, take the gas condensate samples by dry capillary burette and measure the refractive index; then take the liquid samples by another dry capillary burette and measure the refractive index after cleaning the refractometer prism with acetone. After measurement is completed, clean the refractometer prism and release the Ⅰ solution into the original solution bottle (Do not go wrong).

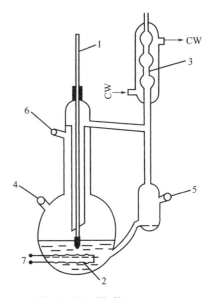

Fig. 2.28　Ebulliometer
1-temperature meter; 2-heating wire; 3-condenser tube;
4-liquid-sampling point; 5-gas condensate-sampling point;
6-to atmosphere; 7-transformer joint

6. Add the Ⅱ solution into the dry ebulliometer and repeat the above steps, then measure the other solution (Ⅲ~Ⅷ) in turn.

7. After accomplishing the experiment, turn off the power of ultra thermostatic water-circulator bath and clean the refractometer prism. Read atmospheric pressure.

Data Analysis

1. Contrast the refractive index of the solution of different labels by the standard curve to find out the corresponding composition of gas and liquid, if the measured temperature does not match with the standard curve, we should adjust it according to that if the temperature increases 1℃ then the refractive index decreases 0.0005.

2. Draw the T-x diagram of ethanol-cyclohexane through the measured temperature against gas and liquid equilibrium compositions, then find out the composition of the minimum constant boiling mixture and the temperature of minimum constant boiling point from the diagram.

3. Calculate the activity coefficients of ethanol-cyclohexane solution which contains ethanol mole fraction of 0.10 and 0.80 at their vapor-liquid equilibrium temperature.

Notes

1. The voltage transformer's output voltage should be slowly raised until the liquid boiling and generally may not exceed 20 V, otherwise it will blow the heating wire.

2. The solution which added to the ebulliometer must make the surface higher than the heating wire inside the glass tube.

3. Each time before the measurement sampling tube and the glass surface of refractometer prism must be blown to dry by rubber suction bulb.

Questions and Further Thoughts

1. Why the ebulliometer and capillary burette must be dry in this experiment? Where do errors of the T-x diagram of this experiment mainly come from?

2. Why the overall composition is the original solution composition in this ebulliometer system? When liquid-gas equilibrium is reached which part is the vapor content and which part is the liquid content?

3. Whether it will affect the experiment when one solution composition takes place a little change?

实验 2.7 电导率的测定及应用

一、实验目的

测定弱电解质溶液的摩尔电导率,从而计算电离度和电离平衡常数。掌握电导率仪的使用方法。

测定强电解质溶液的摩尔电导率,由外推法求算其无限稀释摩尔电导率。

二、实验原理

1-1 型弱电解质在溶液中电离达到平衡时,其平衡常数 $K_c^\ominus$ 与浓度 c 和电离度 α 之间有如下关系:

$$K_c^\ominus = \frac{c\alpha^2}{(1-\alpha)c^\ominus} \tag{2.22}$$

对弱电解质,其电离度 α 等于溶液在浓度为 c 时的摩尔电导率 Λ_m 与溶液在无限稀释时的摩尔电导率 Λ_m^∞ 之比,即:

$$\alpha = \frac{\Lambda_m}{\Lambda_m^\infty} \tag{2.23}$$

将式(2.23)代入式(2.22)可得:

$$K_c^\ominus = \frac{c\Lambda_m^2}{\Lambda_m^\infty (\Lambda_m^\infty - \Lambda_m) c^\ominus} \tag{2.24}$$

或

$$\frac{c}{c^\ominus}\Lambda_m = (\Lambda_m^\infty)^2 K_c^\ominus \frac{1}{\Lambda_m} - \Lambda_m^\infty K_c^\ominus \tag{2.25}$$

式中，Λ_m^∞ 为对给定的物质在温度一定时为常数，其值可据离子的无限稀释摩尔电导率计算：

$$\Lambda_m^\infty = \Lambda_{m,+}^\infty + \Lambda_{m,-}^\infty \tag{2.26}$$

而 Λ_m 与电导率 κ 和溶液浓度有关：

$$\Lambda_m = \frac{K}{c} \tag{2.27}$$

实验测定不同浓度时电解质的电导率 κ，据上式可计算出不同 c 时的摩尔电导率 Λ_m，然后由式（2.23）计算出 α，由式（2.24）计算 $K_c^\ominus$ 或由式（2.25）作 $\frac{c}{c^\ominus}\Lambda_m - \frac{1}{\Lambda_m}$ 图求得 $K_c^\ominus$。

Λ_m^∞（摩尔电导率）随浓度的变化，对强、弱电解质各不相同，对强电解质，科尔劳施（F. Kohlrausch）总结出下列经验式：

$$\Lambda_m = \Lambda_m^\infty - A\sqrt{c} \tag{2.28}$$

式中，A 为经验常数；Λ_m^∞ 为电解质溶液在 $c \approx 0$ 时的摩尔电导率，称为无限稀释摩尔电导率。可见，以 Λ_m 对 $\sqrt{c}$ 作图应得一直线，其外推至纵轴的截距即为 Λ_m^∞。

三、仪器与试剂

DDS-11A 型电导率仪 1 台；恒温槽 1 套；锥形瓶（250mL，100mL）各 2 个；移液管（25mL）3 支。0.1mol·L^{-1} 标准醋酸溶液；0.1 mol·L^{-1} 标准 KCl 溶液。

四、实验步骤

1. 打开电导率仪器（DDS-11A 型电导率仪的使用参见"4.7DDS-ⅡA 型电导率仪"）。

2. 用 25mL 移液管取 50mL 0.1mol·L^{-1} 标准醋酸溶液放入恒温槽内的 100mL 锥形瓶中，恒温 15min，测定溶液的电导率。不时摇动，至三次读数接近为止。

3. 用吸取醋酸的移液管从已测溶液中吸出 25mL 弃去。用另一支移液管取 25mL 电导水注入已吸出溶液的 100mL 锥形瓶中，同法测定其电导率，如此稀释四次测定其电导率。

4. 倒去所测醋酸溶液，洗净锥形瓶，并用电导水清洗、浸泡电极数分钟，然后取 50mL 电导水测其电导。

5. 调整恒温槽温度在（25.0±0.1）℃（室温高时，可调至 30℃）。

6. 在广口瓶中，用 25mL 移液管加入 50mL 0.1mol·L^{-1} 标准 KCl 溶液，插入电极，然后置于恒温水浴中。等 15～20min 至温度恒定后，测其电导率 κ，至三次读数接近为止。

7. 用吸取 KCl 溶液的移液管从电导池中吸出 25mL 溶液弃去，用取水的移液管移取 25mL 电导水注入电导池中，混合均匀，待温度恒定后测其电导率 κ，如此稀释四次，分别测其电导率 κ。实验数据记录见表 2.14。

实验完毕，关上电源，倒掉锥形瓶中的电导水，取出电极，拆卸装置。

五、数据处理

1. 计算各浓度醋酸的电导率。醋酸的电导率很小，其值等于所测溶液的电导率减去同温度电导水的电导率。

2. 计算醋酸在各浓度下的摩尔电导率 Λ_m^∞ 和电离度 α。已知 25℃时醋酸的 $\Lambda_m^\infty = 3.907 \times 10^{-2}$ S·m^2·mol^{-1}。

3. 从各 α 值计算 $K_c^\ominus$，并求其平均值。

表 2.14 电导率测定数据记录

电极常数：　　　　　　　　实验温度：

HAc 浓度 /mol·L^{-1}	次数	溶液电导率 /S·m^{-1}	醋酸电导率 /S·m^{-1}	摩尔电导率 /S·m^2·mol^{-1}	电离度 α	电离常数 $K_c^{\ominus}$
	1					
	2					
	3					
	1					
	2					
	3					
	1					
	2					
	3					
	1					
	2					
	3					
电导水	1					
	2					
	3					

4. 以 $\dfrac{c}{c^{\ominus}}\Lambda_m$ 对 $\dfrac{1}{\Lambda_m}$ 作图，由直线的斜率求出 $K_c^{\ominus}$。

5. 计算氯化钾各浓度溶液的电导率。

6. 以 KCl 溶液的 Λ_m 对 $\sqrt{c}$ 作图，外推求 Λ_m^{∞}，并与文献值比较。

7. 写出 KCl 溶液的摩尔电导率与浓度的具体关系式。

$$\Lambda_m = \Lambda_m^{\infty} - A\sqrt{c}$$

将上述结果与文献值比较。

六、注意事项

1. 电导率仪电源连接为五芯插头，切勿插错以免电极烧毁。
2. 严格按照仪器说明书的使用方法操作仪器。
3. 每次测量之前都要校正，待稳定时方可读数。
4. 电极引线不能受潮，以免短路测不准确。
5. 若用滤纸拭干电极，切勿触及铂黑。
6. 移液管不能混用。

七、思考题

1. 什么叫溶液的电导、电导率和摩尔电导率？
2. 影响摩尔电导率的因素有哪些？
3. 如何测定电导池常数？
4. 测定电导率为何要使用交流电源？
5. 强、弱电解质溶液的摩尔电导率与浓度的关系有何不同？
6. 为什么强电解质的摩尔电导率随溶液浓度的减小而增大？

Experiment 2.7 Determination and Applications of Electrical Conductance

2.7.1 Ionization Constant of a Weak Acid

Objectives

1. To determine the molar conductivtity of weak electrolyte and to calculate the degree and the equilibrium constant of ionization.

2. To learn the operation of a conductivity meter.

Theory/Principle

When there exists an ionization equilibrium for a AB-type of weak electrolyte in solution, the relationship between the ionization constant $K_c^\ominus$ and the degree of ionization α and the concentration c is

$$K_c^\ominus = \frac{c\alpha^2}{(1-\alpha) \ c^\ominus} \tag{2.22}$$

For a weak electrolyte, the ratio of molar conductivity Λ_m at a specified concentration c to the limiting molar conductivity Λ_m^∞ represents the degree of ionization α:

$$\alpha = \frac{\Lambda_m}{\Lambda_m^\infty} \tag{2.23}$$

When α in equation (2.22) is replaced by ratio of $\Lambda_m \sim \Lambda_m^\infty$, we can get the following equation:

$$K_c^\ominus = \frac{c\Lambda_m^2}{\Lambda_m^\infty (\Lambda_m^\infty - \Lambda_m) \ c^\ominus} \tag{2.24}$$

or

$$\frac{c}{c^\ominus}\Lambda_m = (\Lambda_m^\infty)^2 K_c^\ominus \frac{1}{\Lambda_m} - \Lambda_m^\infty K_c^\ominus \tag{2.25}$$

The limiting molar conductivity Λ_m^∞ of a given solute is a constant at a specific temperature, and it can be calculated from the limiting molar conductivity Λ_m^∞ of ions contained in the solute:

$$\Lambda_m^\infty = \Lambda_{m,+}^\infty + \Lambda_{m,-}^\infty \tag{2.26}$$

And the molar conductivity Λ_m is related to the specific conductance K and the concentration of the solution:

$$\Lambda_m = \frac{K}{c} \tag{2.27}$$

In this experiment, the conductivity K of solute at different concentrations will be determined, and the molar conductivity Λ_m at different concentrations can be calculated from the equation (2.27). Then we can calculate the ionization degree from equation (2.23) and the dissociation constant $K_c^\ominus$ from equation (2.24) or the plot of $\frac{c}{c^\ominus}\Lambda_m$ against $\frac{1}{\Lambda_m}$ according to the equation (2.25). DDS-11A conductivity meter, whose operating instruction is detailed in the appendix, is used to measure the conductivity K.

Apparatus and Chemicals Required

Apparatus: DDS-11A conductivity meter, electrode, thermostatic bath, 250mL and 100mL conical flask, 25mL transfer pipet.

Chemicals: 0.1mol/L standard acetic acid solution.

Procedure

1. Transfer 50 mL of 0.1 mol/L standard acetic acid solution with 25 mL pipette to a 100 mL conical flask in the thermostatic bath. After 15 minutes, a stable temperature will be reached, and measure the conductivity. Shake the conical flask and repeat the measurement for three times.

2. Withdraw 25 mL solution from the solution, which has been determined, with the pipette used to transfer acetic acid solution and discard it. Transfer 25 mL conducting water with another pipette into the 100 mL conical flask that has been thrown away 25 mL solution. Determine the conductivities as procedure 1. Dilute the solution for four times following the above steps and measure the conductivities.

3. Throw away the measured acetic acid solution and clean the conical flask with conducting water. Clean and immerse the electrodes for serval minutes. Then determine the conductivity of 50 mL conducting water.

4. When the experiment is finished, shut the power supply, throw away the conductance water in the conical flask and take out the electrodes, disassemble the equipment.

Observations and Measurements

Table 2.15a Conductivities of the HAc solution and the conductance water

constant of electrodes _____ room temperature _____ ℃ atmospheric pressure _____ Pa

Concentration of HAc (c)/mol·L^{-1}	times	Conductivity of solution (K)/S·m^{-1}	Conductivity of HAc /S·m^{-1}	times	Molar Conductivity /S·m^2·mol^{-1}	ionization degree	ionization constant
	1			1			
	2			2			
	3			3			
	1			1			
	2			2			
	3			3			
	1			1			
	2			2			
	3			3			
Conducting water	1						
	2						
	3						

Calculations

1. Calculate the conductivities of HAc at each concentration. Since the conductivity of HAc is minor, its value equals to the difference of the conductivities of the measured solution and the conducting water at the same temperature.

2. Calculate the molar conductivity Λ_m and the degree of ionization α of HAc at each concentration. The limiting molar conductivity of HAc at the temperature of 298 K is 0.03907 S·m²·mol⁻¹.

3. Calculate the dissociation constant $K_c^\ominus$ according to each degree of ionization α, then work out its mean value.

4. Plot $\dfrac{c}{c^\ominus}\Lambda_m$ against $\dfrac{1}{\Lambda_m}$, and calculate the dissociation constant $K_c^\ominus$ from the line slope.

5. Compare the above value of $K_c^\ominus$ with that recorded in the literature.

Notes

1. The power connection of the conductivity meter is five-pin plug; do not misplug the electrode or the electrode will be burned out.

2. Operate the apparatus according to the operation method described in the appendix.

3. Make correction before measurement, and record the data when it gets stable.

4. Do not wet the electrode wire or you will get short circuit and inaccurate data.

5. Take care not to touch the platinum black when you wipe the electrode with filter paper.

6. Do not put off the transfer pipette.

Questions/Exercises

1. What is conductance, conductivity and molar conductivity of solution?

2. What factors will influence the molar conductivity?

3. How to determine the cell constant?

4. Why should we use alternating current when measuring the conductivity?

Appendix

Operation of DDS-11A type conductivity meter:

Preheat the instrument for more than 10min, and check whether there is a full scale display. Choose the cell constant according to the conductance electrode used. Select the working frequency. Choose "high frequency", if the conductivity is higher than 10^3, otherwise choose "low frequency". Set the switch to position of "calibration", and turn "adjust" to have a full scale. Then set the switch to "measurement", read the display and record it. The selection of measurement range should be from large to small. Pay attention to the display difference between the measurement range in red and measurement range in black.

2.7.2 Conductance Behavior of Strong Electrolytes

Objectives

To determine the molar conductivtity of strong electrolyte and to calculate the limiting molar conductivity by using extrapolation method.

Theory/Principle

The relationship between the molar conductivity Λ_m and the concentration is different for weak electrolytes and strong electrolytes. For strong electrolytes in low-concentration range, the F. Kohlrausch equation holds:

$$\Lambda_m = \Lambda_m^\infty - A\sqrt{c} \tag{2.28}$$

Where Λ_m is molar conductivity at concentration c; Λ_m^∞ is molar conductivity at infinite dilution; c is concentration.

Obviously we can take advantage of the linear form of the relationship to determine Λ_m^∞ graphically. Experimental values of Λ_m are plotted versus $\sqrt{c}$ and the graph is extrapolated back to zero concentration which corresponds to Λ_m^∞.

The molar conductivity Λ_m depends on the specific conductance K and the concentration of the solution:

$$\Lambda_m = \frac{K}{c} \tag{2.27}$$

In this experiment the conductivities at different concentration will be determined directly, and the corresponding molar conductivities will be calculated according to equation (2.27).

Apparatus and Chemicals Required

Apparatus: DDS-11A conductivity meter, thermostatic bath, 125 mL conical flask, 25 mL transfer pipet.

Chemicals: 0.1 mol/L standard potassium chloride solution.

Procedure

1. Turn on the power of the thermostatic bath, and adjust the temperature to $25.0\ ℃ \pm 0.1\ ℃$. When the room temperature is high, we can adjust the temperature to $30.0\ ℃$.

2. Transfer 50 mL of $0.1\ mol \cdot L^{-1}$ standard potassium chloride solution with 25mL pipette to a conical flask, plug the electrodes in, and put the flask into the thermostatic bath. After about 15~20 minutes, a stable temperature will be reached, and measure the conductivity. Shake the conical flask and repeat the measurement for three times.

3. Remove 25 mL solution from the cell with the pipettes used to transfer potassium chloride solution and discard it. Transfer 25 mL conducting water with another pipette into the conductance cell. Shake the flask to mix the solution properly. Determine the conductivities when the termperature is stable. Dilute the solution for four times following the above steps and measure the conductivities.

Observations and Measurements

Table 2.15b Conductivities of the KCl solution

constant of electrodes _____ room temperature _____ ℃ atmospheric pressure _____ Pa

Concentration of KCl (c)/mol $\cdot$ L^{-1}	Times	Conductivity of solution(K)/S $\cdot$ m^{-1}	Molar Conductivity of KCl /S $\cdot$ m^2 $\cdot$ mol^{-1}
	1		
	2		
	3		
	1		
	2		
	3		
	1		
	2		
	3		

Calculations

1. Calculate the molar conductivities at each concentration.

2. Plot Λ_m against $\sqrt{c}$, and calculate Λ_m^∞ by using extrapolation method. Then compare the above value of Λ_m^∞ with the literature value.

3. Give out the specific relational expression between Λ_m and $\sqrt{c}$ of potassium chloride solution: $\Lambda_m = \Lambda_m^\infty - A\sqrt{c}$.

Questions/Exercises

1. What is the difference of the relationship between the molar conductivity Λ_m and the concentration for weak electrolytes and strong electrolytes?

2. Why does the molar conductivity increase as the solution is diluted for strong electrolytes?

实验 2.8 电动势的测定及其应用

一、实验目的

1. 掌握可逆电池电动势的测量原理及电位差计的使用方法。
2. 学会构成可逆电池及测定几个原电池电动势。

二、实验原理

热力学可逆电池必须具备两个条件：①电池中各电极反应是可以逆向进行的；②必须在非常接近平衡状态的条件下工作（通过的电流无限小）。

可逆电池电动势等于电池两个电极电势之差。

$$E = E_+ - E_- \tag{2.29}$$

E_+、E_- 分别为正、负极的电极电势。它们与参加电极反应各物质的活度之间服从能斯特方程。对任意一个给定电极，其电极反应通式为：

$$\text{氧化态} + ze \Longrightarrow \text{还原态}$$

z 为进行上述电极反应所需电子的物质的量。则电极电势 E 的通式为：

$$E = E^\ominus - \frac{RT}{zF} \ln \frac{a_{\text{还原态}}}{a_{\text{氧化态}}} \tag{2.30}$$

式中，$E^\ominus$ 为该给定电极的标准电极电势。

由于可逆电池过程必须在电流趋于零的条件下进行，所以，可逆电池电动势的测定必须用补偿法。补偿法测电位差所用的仪器称为电位差计。本实验使用 UJ-25 型高电势直流电位差计测定可逆电池电动势，其原理及使用方法参见"4.8 补偿法原理及 UJ-25 型高电势直流电位差计"。

本实验测定 3 种可逆电池电动势，现分述如下。

1. 银电极与甘汞电极构成的可逆电池

$$\text{Hg，Hg}_2\text{Cl}_2(s) \mid \text{饱和 KCl 溶液} \parallel \text{AgNO}_3(a) \mid \text{Ag}$$

其电池电动势为：
$$E = E_{Ag^+/Ag} - E_{饱和甘汞} \tag{2.31}$$

饱和甘汞电极的 $E_{饱和甘汞}$ 与 t（℃）关系为：
$$E_{饱和甘汞}/V = 0.2415 - 7.6 \times 10^{-4}(t/℃ - 25) \tag{2.32}$$

由实验可测得电池电动势 E，$E_{饱和甘汞}$ 可由式（2.32）求得，因此可求得 $E_{Ag^+/Ag}$。

2. 铜电极与饱和甘汞电极构成的可逆电池

$$Hg，Hg_2Cl_2(s) | 饱和 KCl 溶液 \| CuSO_4(a) | Cu$$

其电池电动势为：
$$E = E_{Cu^{2+}/Cu} - E_{饱和甘汞} \tag{2.33}$$

测得电动势 E 即可求出 $E_{Cu^{2+}/Cu}$。

3. 醌氢醌电极与甘汞电极构成的可逆电池

醌氢醌电极是一种氢离子电极，主要用于溶液 pH 值的测定。通常，待测溶液构成醌氢醌电极，和参比电极（一般为甘汞电极）相连构成电池，测定电动势，进而求出所测溶液的 pH 值。

醌氢醌是一种醌和氢醌的等摩尔比的复合物，为深褐色固体粉末，微溶于水。被溶解部分能完全分解为等量的醌（Q）和氢醌（H_2Q）。因此只要将少量的醌氢醌加入待测溶液中，形成醌氢醌的饱和溶液，插入一个铂电极，即构成醌氢醌电极。其电极反应为：

$$Q + 2H^+ + 2e^- \rightleftharpoons H_2Q$$

$$E_{Q \cdot H_2Q} = E^{\ominus}_{Q \cdot H_2Q} - \frac{RT}{2F} \ln \frac{a_{H_2Q}}{a_{H^+} a_Q}$$

由于醌、氢醌在溶液中浓度相等且很低，故可视为 $a_Q = a_{H_2Q}$，则

$$E_{Q \cdot H_2Q} = E^{\ominus}_{Q \cdot H_2Q} - \frac{RT}{F} \ln \frac{1}{a_{H^+}}$$

$$= E^{\ominus}_{Q \cdot H_2Q} - \frac{2.303RT}{F} pH \tag{2.34}$$

将醌氢醌电极与甘汞电极组成电池为：

$$Hg，Hg_2Cl_2(s) | 饱和 KCl 溶液 \| 待测 pH 溶液，Q \cdot H_2Q | Pt$$

电动势为：
$$E = E_{Q \cdot H_2Q} - E_{饱和甘汞}$$

将式（2.33）代入得：
$$pH = \frac{E^{\ominus}_{Q \cdot H_2Q} - E_{饱和甘汞} - E}{2.303RT/F} \tag{2.35}$$

其中 $E^{\ominus}_{Q \cdot H_2Q}$ 与温度 t（℃）的关系为：
$$E^{\ominus}_{Q \cdot H_2Q}/V = 0.6994 - 7.4 \times 10^{-4}(t/℃ - 25) \tag{2.36}$$

应当注意，醌氢醌电极不能用于碱性溶液中。当 pH > 8.5 时，由于氢醌大量电离，影响其浓度，使 $a_Q = a_{H_2Q}$ 不能成立，从而影响测定结果。

三、仪器与试剂

SDC 数字电位差综合测试仪或 UJ-25 型电位差计，检流计，工作电池（2个），饱和甘汞电极，铂电极，银电极，铜电极，小烧杯，盐桥，醌氢醌，饱和 KCl 溶液，0.100 mol·L^{-1} 的 $AgNO_3$ 溶液，0.100 mol·L^{-1} 的 $CuSO_4$ 溶液，NaAc 与 HAc 配制的缓冲溶液（pH

溶液）。

四、实验步骤

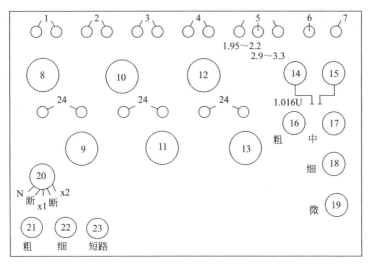

图 2.29　UJ-25 型电位差计板面布置图

1. UJ-25 型电位差计测电位差 (图 2.29) 的使用方法

（1）线路连接　将转换开关放在"断"位置，将左上方的 4 个按钮全部松开，分别连接检流计、工作电池、标准电池及待测电池。接各电池时注意正、负极位置。

（2）调整工作电流　首先将标准电池温度补偿 A、B 调到标准电池的电动势值（此值由标准电池电动势与温度关系式，按实验时的室温算出）。然后将转换开关 K 放在 N 位置上，按下"粗"按钮，使高电阻 r 接入测量电路，避免检流计和标准电池中电流过大而受损伤。检流计光点发生偏转，立即松开按钮。根据光点偏转方向由粗至细地调可变电阻 R。反复多次按"粗"按钮及调节，直至检流计光点偏转不明显。再按"细"按钮调可变电阻值，直至检流计光点不偏转为止。此时工作电流为 0.0001A。调节中，松开"粗"、"细"按钮时若检流计光点摆动不停，可按"短路"按钮，使光点较快停在零点。

（3）测量待测电池电动势　将转换开关 K 放在"x1"或"x2"位置上。首先按"粗"按钮，调节第 1～3 测量盘旋钮，使检流计光点偏转不明显。再按"细"按钮，调节 4～6 测量盘旋钮，使检流计光点不偏转为止。从 6 个读数窗孔内读出待测电池电动势的各位数字。

2. 标定工作电流

接好检流计、标准电池、工作电池（甲电池）。经教师检查后，将电位差计的旋钮 14 及 15 的指示值调到标准电池的电动势值。其值与温度 t（℃）的关系为：

$$E_{标准}/V = 1.0186 - 4.06 \times 10^{-5} (t/℃ - 20) - 9.5 \times 10^{-7} (t/℃ - 20)^2 \quad (2.37)$$

然后，将转向开关 20 放在"N"位置上。按下粗调按钮 21，检流计光点发生偏转，立即松开按钮 14，根据光点偏转方向，调节可变电阻 16、17、18。再按按钮，进行反复调节，将光点偏转调至很小，再按细调按钮 22，调节可变电阻 18、19，直至检流计光点不再偏转为止。工作电流标定完毕，可变电阻 16、17、18、19 的阻值不能再改变。

注意：每次按按钮的时间不应太长，看到检流计光点偏转方向立即松开，以使电流通过标准电池的时间尽量缩短。

3. SDC 数字电位差综合测试仪的使用方法

（1）开机　用电源线将仪表后面板的电源插座与 220V 电源连接，打开电源开关（ON），预热 15min。

（2）以内标为基准进行测量

① 校验

a. 用测试线将被测电动势按"＋"、"－"极性与测量插孔连接。

b. 将"测量选择"旋钮置于"内标"。

c. 将"10^0"位旋钮置于"1"，补偿旋钮逆时针旋到底，其他旋钮均置于"0"，此时，"电位指示"显示"1.00000"V。

② 测量

a. 将"测量选择"置于"测量"。

b. 调节"$10^0 \sim 10^{-4}$"五个旋钮，使"检零指示"显示数值为负且绝对值最小。

c. 调节"补偿旋钮"，使"检零指示"显示为"0000"，此时"电位显示"数值即为被测电动势的值。

（3）以外标为基准进行测量

① 校验

a. 将已知电动势的标准电池按"＋"、"－"极性与"外标插孔"连接。

b. 将"测量选择"旋钮置于"外标"。

c. 调节"$10^0 \sim 10^{-4}$"五个旋钮和"补偿"旋钮，使"电位指示"显示的数值与外标电池数值相同。

d. 待"检零指示"数值稳定后，按一下采零键，此时，"检零指示"显示为"0000"。

② 测量

a. 拔出"外标插孔"的测试线。再用测试线将被测电动势按"＋"、"－"极性接入"测量插孔"。

b. 将"测量选择"置于"测量"。

c. 调节"$10^0 \sim 10^{-4}$"五个旋钮，使"检零指示"显示数值为负且绝对值最小。

d. 调节"补偿旋钮"，使"检零指示"显示为"0000"，此时"电位显示"数值即为被测电动势的值。

（4）关机　首先关闭电源（OFF），然后拔下电源线。

4. 组成电池

按原理中所述组成三组电池。其中负极均为饱和甘汞电极，其构成为：向一小烧杯中倒入 30mL 饱和 KCl 溶液，插入甘汞电极即为负极；正极分别由三种不同的溶液加不同的电极构成。分别为：

① 30mL pH 溶液，加入少量（约 0.1g）醌氢醌，插入铂电极；

② 30mL 0.100 mol·L^{-1} 的 $CuSO_4$ 溶液，插入新镀铜的铜电极；

③ 30mL 0.100 mol·L^{-1} 的 $AgNO_3$ 溶液，插入银电极。

每组电池间以盐桥连接以消除液接电势，盐桥使用前以蒸馏水淋洗，并用滤纸吸干水分。将待测电池的正负极对应接到电位差计的测量柱上。

5. 测电池电动势

将转向开关 20 放在与电池对应的"x1"或"x2"位置上。首先按"粗"按钮，调节旋钮 8、9、10、11。使检流计光点无明显偏转。再按"细"按钮，调节旋钮 11、12、13，使检流计光点不偏转为止。由 6 个读数窗孔读出被测电池电动势的各位数字。重复两次，取平均值。

五、 数据处理

1. 写出各被测电池反应和电池反应。
2. 计算 pH 溶液的 pH 值；计算铜电极的 $E_{Cu^{2+}/Cu}$；计算银电极的 $E_{Ag^+/Ag}$。

六、 思考题

1. 本实验的测定过程中，在找到光点不偏转的位置之前，仍有电流通过被测电池，对测量会带来什么影响？如何减少这种影响？
2. 试分析检流计光点只往一边偏转，而找不到光点不发生偏转的平衡点，可能的原因是什么？
3. 实际测量过程中，用的是标准电池还是工作电池？为什么？

Experiment 2.8　　Determination of Electromotive Force of Reversible Cell

Objectives

1. To understand the measurement principle of electromotive force and the application method of the potentiometer.
2. To grasp the structure of reversible cells and to measure the electromotive force of cells.

Theory and Principle

Thermodynamic reversible cells must have two conditions: (1) electrode reactions in cell could be reversible; (2) electrode reactions must be work under balancing condition (the little current in the wire).

The potentials of reversible cell may be calculated from the equation:

$$E = E_+ - E_- \tag{2.29}$$

Where E is the cell potential; E_+ is the anode electrode potential; E_- is the cathode electrode potential.

The single potential between a oxidation and a reduction may be calculated from the Nernst equation:

$$\text{oxidation} + ze \rightleftharpoons \text{reduction}$$

and the emf is:

$$E = E^\ominus - \frac{RT}{zF} \ln \frac{a_{\text{reduction}}}{a_{\text{oxidation}}} \tag{2.30}$$

Where $E^\ominus$ is the standard reduction potential; R is the gas constant 8.316 joules/degree; n is the valence of the ion; F is the value of the Faraday 96500 coulombs; a is the activity of oxidation or reduction

Since the reversible cell works under the condition of zero current, the potential of reversible cell was measured by the method of compensation, the apparatus used in this experiment was called potentiometer. In this experiment UJ-25 type potentiometer was used to measure the potential of reversible cell. Its theory and applying method was illustrated in appendix eight.

Three kinds of reversible cells were used in this experiment. They were introduced according to nest.

1. The reversible cell structured by the silver electrode and the calomel electrode:
Set up the cell
$$\text{Hg, } Hg_2Cl_2(s) \mid \text{saturated KCl} \parallel AgNO_3(a) \mid Ag$$
The emf of the cell is
$$E = E_{Ag^+/Ag} - E_{\text{calomel electrode}} \tag{2.31}$$
The potential of saturated calomel electrode is a function of temperature:
$$E_{\text{calomel electrode}}/V = 0.2415 - 7.6 \times 10^{-4}(t/^\circ C - 25) \tag{2.32}$$

The E of cell can be obtained by experiment, and the potential of saturated calomel electrode would be calculated according to the (2.32) equation. Therefore the potential of silver electrode was obtained.

2. The reversible cell set up by copper electrode and saturated calomel electrode:
Set up the cell
$$\text{Hg, } Hg_2Cl_2(s) \mid \text{saturated KCl} \parallel CuSO_4(a) \mid Cu$$
The emf of the cell is
$$E = E_{Cu^{2+}/Cu} - E_{\text{saturated calomel}} \tag{2.33}$$

The E can be measured by experiment, therefore the potential of copper electrode can be calculated.

3. The reversible cell made up by the quinhydrone electrode and saturated calomel electrode.

The quinhydrone electrode is now only of historical interest. However, it is a good example of a chemical oxidation reduction system sensitive to pH, which can be made into a pH electrode. The material, quinhydrone, $C_6H_4O_2 \cdot C_6H_4(OH)_2$, in aqueous solution dissociates into equal numbers of quinone, $C_6H_4O_2$, and hydroquinone, $C_6H_4(OH)_2$, molecules. The quinine and hydroquinone molecules set up the oxidation-reduction system.

$$C_6H_4O_2 + 2H_3O^+ + 2e \rightleftharpoons C_6H_4(OH)_2 + 2H_2O$$

The potential of this half-cell reaction is
$$E_{Q \cdot H_2Q} = E^{\ominus}_{Q \cdot H_2Q} - \frac{RT}{2F} \ln \frac{a_{H_2Q}}{a_{H^+} a_Q}$$

Where Q is quinine; H_2Q is hydroquinone.

Since the concentrations of quinine and hydroquinone are the same and the activity coefficient of these non-electrolytes are approximately equal, the equation reduces to
$$E_{Q \cdot H_2Q} = E^{\ominus}_{Q \cdot H_2Q} - \frac{RT}{F} \ln \frac{1}{a_{H^+}}$$
$$= E^{\ominus}_{Q \cdot H_2Q} - \frac{2.303RT}{F} pH \tag{2.34}$$

4. Set up the cell with the quinhydrone electrode and saturated calomel electrode

$$\text{Hg, Hg}_2\text{Cl}_2 \text{ (s)} \mid \text{saturated KCl} \parallel \text{pH solution, Q} \cdot \text{H}_2\text{Q} \mid \text{Pt}$$

The emf of the cell is

$$E = E_{\text{Q} \cdot \text{H}_2\text{Q}} - E_{\text{saturated calomel}}$$

The equation (2.34) is rewritten as

$$\text{pH} = \frac{E^{\ominus}_{\text{Q} \cdot \text{H}_2\text{Q}} - E_{\text{saturated calomel}} - E}{2.303RT/F} \tag{2.35}$$

The potential of quinhydrone electrode is a function of temperature

$$E^{\ominus}_{\text{Q} \cdot \text{H}_2\text{Q}}/V = 0.6994 - 7.4 \times 10^{-4} (t/℃ - 25) \tag{2.36}$$

Where pH is defined as the negative logarithm of the hydrogen ion activity. The quinhydrone electrode can not be used at pH's above 8 because of oxidation by atmospheric oxygen and the ionization of the weakly acidic hydroquinone.

Apparatus and Chemicals Required

SDC digital potentiometer or UJ-25 potentiometer (Fig. 2.30); galvanometer; two working batteries; saturated calomel electrode; Pt electrode; silver electrode; copper electrode; litter beaker; salt bridge; quinhydrone; saturated KCl solution; 0.1 mol·L^{-1} AgNO$_3$ solution; 0.1 mol·L^{-1} CuSO$_4$ solution; pH solution made from NaAc and HAc.

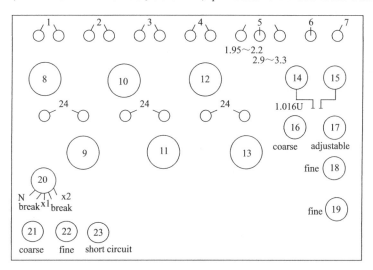

Fig. 2.30 The figure of the structure of UJ-25 potentiometer

Method of Procedure

1. The applying method of UJ-25 potentiometer

(1) Connect wire Set the change-over key on the place of the "break", the four button which line on the left top were loosened. Connect the galvanometer, work battery, standard cell and waiting measured cell, then call for the "position" of negative and positive.

(2) Adjust work current At first, set the button A、B of the standard cell for the value corresponding to the known E. M. F. of the standard cell (the emf of the standard can be calculated according to the equation of between potential and temperature under the condition of experiment). Then set the K key on the "N" position, quickly tap the "Coarse adjust-

ment" button, and note the deflection of the galvanometer. Adjust the regulating rheostat R until no deflection of the galvanometer is noticed when "Coarse adjustment" button is tapped. By the way, adjust the regulating rheostat R until no deflection of the galvanometer is noticed when "fine adjustment" button is tapped. (Note: the key "adjustment" should never hold down, as a current flow of any magnitude through the standard cell will cause its potential to rise or fall, through polarization.)

(3) Determination of electromotive force   Set key K on the position of x1 or x2, at first adjust the 1~3 dials until no deflection of the galvanometer is noticed when the "Coarse adjustment" is tapped; then adjust the 4~6 dials until no deflection of the galvanometer is noticed when the "fine adjustment" is tapped. The electromotive force then are obtained from 1~6 dials.

2. Calibrate the working current flow

Connect the galvanometer, standard cell and working battery. After looked through by teacher, set the 14 and 15 buttons (see appendix 8) of the potentiometer for the value corresponding to the known E.M.F. of the standard cell. The emf of standard cell is a function of temperature:

$$E/V = 1.0186 - 4.06 \times 10^{-5}(t/℃ - 20) - 9.5 \times 10^{-7}(t/℃ - 20)^2 \tag{2.37}$$

Then set the key 20 on the "N" position. Tap down the 21 button, and there was a deflection in the galvanometer. Adjust the regulating rheostat 16, 17, 18 until no deflection of the galvanometer is noticed when button 21 is tapped down. By the way, adjust the regulating rheostat 18, 19 until no deflection of the galvanometer is noticed when button 22 is tapped down. After completing the work of calibrating working battery, this regulating rheostats 16, 17, 18, 19 would not be adjusted again.

Notice: when there is a deflection in the galvanometer, buttons 21, 22 should never be held down a long time in order to reduce the time of a current through standard cell.

3. The applying method of SDC digital potentiometer

(1) Turn on   Connect the instruments and 220V current source with wire, turn on the key of current, and hold up for 15 minutes.

(2) Measurement of electromotive force on the internal standard

① Calibrate

a. Connect the "+" or "−" polar of measured cell to the jack of measurement with wire.

b. Set the button of "measurement choice" on the position of "internal standard".

c. Throw "10^0" into the "1" position. The button of compensation goes round counterclockwise to the end, other buttons was set the position of "0", and "1.00000 V" will be happen on the potentiometer. Tap the "picks zero key" to get "0000" on the screen of "zero current indicator".

② Measurement

a. Throw the "measurement choice" into the "measurement" position.

b. Adjust "$10^0 \sim 10^{-4}$" five knobs, until the demonstration value is negative on the "zero current indicator", and the absolute value is smallest.

c. Adjust the "compensates knob", until the demonstration is "0000" on the "zero current indicator". This time, the value is the electromotive force.

(3) Measurement of electromotive force on the external standard

① Calibrate

a. Connect the "+", "−" polar of the standard battery and "the external standard jack".

b. Put The "choice knob" in the "external standard".

c. Adjust "$10^0 \sim 10^{-4}$" five knobs, until the demonstration value is the value of the external standard cell.

d. After the demonstration value is stable, tap the "picks zero key". This time, the demonstration value is "0000".

② Measurement

a. Draw out the slotted line of "external standard jack", and then connect "+", "−" polar of the measured electromotive force with the "survey jack" using the slotted line.

b. Set the button of " measurement choice" on the position of "measurement".

c. Adjust "$10^0 \sim 10^{-4}$" five knobs, until the demonstration value is negative, and the absolute value is smallest.

d. Adjust "compensates the knob", until the demonstration value is "0000". This time, the electric potential value of potentiometer is the electromotive force.

(4) Turn off First close the power source (OFF), then tear off the power line.

4. Set up the cell

Set up the three groups of cells according to the stated principle, and the saturated calomel electrode is used as cathode. The saturated KCl solution is poured into a 30 mL small beaker, the inserted calomel electrode is the cathode. The positive electrode constituted separately by three kind of different solutions respectively is:

① 30mL pH solution / (0.1g) hydroquinone / Pt;

② 30mL 0.1 mol/L $CuSO_4$ solution / Cu;

③ 30ml 0.1 mol/L $AgNO_3$ solution / Ag.

The electric potential of liquid junction potential would be eliminated by the salt bridge. Before the salt bridge used, it would be washed by the distilled water and sucked with the filter paper. The tested battery's positive and negative polar were connected on the column of the potentiometer.

5. Measure the battery electromotive force

Throw the key 20 into the "x1" or "x2" position. At first, adjust knob 8, 9, 10, 11 until no obviously deflection in the galvanometer is noticed when the "coarse adjustment" button is pressed. Press the "fine adjustment" button again, adjust knob 11, 12, 13, until the galvanometer luminous spot will deflect. The value of battery electromotive force can be obtained from the six reading opening each digit. Repeat the method for two times and average the value.

Calculations

1. Write the measured cell reaction respectively;

2. Calculate the pH of the pH solution, the electromotive force of the copper electrode and silver electrode.

Questions and Exercises

1. In this experiment's determination process, before finding the position which the luminous spot did not deflect, still had the electric current through to measure the battery, what effect can it bring to the surrey? How to reduce this kind of influence?

2. Try to analyze the galvanometer luminous spot only to deflect toward one side, but could not find the luminous spot not to have the deflection balance point, what was the possible reason? In actual measuring process, what we used is the standard battery or the working battery? Why?

实验2.9 蔗糖水解速率常数的测定

一、实验目的

根据物质的光学性质研究蔗糖水解反应，测定其反应速率常数和半衰期，了解旋光仪的基本原理，掌握旋光仪的使用方法。

二、实验原理

蔗糖在水中水解成葡萄糖和果糖的反应式为（H^+起催化作用）：

$$C_{12}H_{22}O_{11} + H_2O \xrightarrow{H^+} C_6H_{12}O_6 + C_6H_{12}O_6$$

这是个二级反应，对蔗糖为一级，对水也为一级。由于水过量其浓度变化甚微可视为常数，故该反应可按一级反应处理（称为准一级反应），其反应速率方程的微分式和积分式分别表示为：

$$-\frac{dc_A}{dt} = kc_A \tag{2.38}$$

$$\ln c_A = -kt + \ln c_{A,0} \tag{2.39}$$

式中，k为反应速率常数；$c_{A,0}$为蔗糖初浓度；c_A为反应进行到任意时刻t时蔗糖的浓度；t为反应时间。

当$c_A = c_{A,0}/2$时所需的反应时间为反应的半衰期，用$t_{1/2}$表示：

$$t_{1/2} = \frac{\ln 2}{k} \tag{2.40}$$

如果实验测得不同反应时刻（t）蔗糖的浓度（c_A），作$\ln c_A$-t图为直线，由其斜率可求得k，进而可求$t_{1/2}$。

在本反应中，反应物和产物都具有旋光性（含有不对称的碳原子），且旋光能力不同，故可用反应系统的水解过程中旋光性质的变化来度量反应进度。

物质的旋光性是指该物质可以使通过其中的一束偏振光的偏振面旋转某一角度的性质，具有此种性质的物质称为旋光性物质。物质的旋光能力以其使偏振面旋转的角度来度量，该角度称为旋光度，对含有旋光性物质的溶液，其旋光度与旋光物质的本性、溶剂性质、入射光的波长、样品管长度、溶液浓度和温度等因素有关。当物性、入射光波长及温度一定时，

溶液旋光度 α 与浓度、样品管长度成正比

$$\alpha = [\alpha]_\lambda^t \frac{lc}{100} \tag{2.41}$$

式中，l 为样品管长度，dm；c 为溶液浓度，g/100mL；$[\alpha]_\lambda^t$ 为比例常数，λ 表示偏振光波长，t 表示摄氏温度。其物理意义为：当波长为 λ 的偏振光通过 1dm 厚、每 1mL 中含有 1g 旋光物质的溶液时所产生的旋光度，亦称为比旋光度。可用比旋光度比较各种旋光性物质的旋光能力。本实验中，反应物和产物的比旋光度分别为：

蔗糖　　　　$[\alpha]_D^{20} = 66.65°$

葡萄糖　　　$[\alpha]_D^{20} = 52.5°$

果糖　　　　$[\alpha]_D^{20} = -90.9°$

$[\alpha]_D^{20}$ 的下标表示偏振光为钠黄光（$\lambda=589$nm），$[\alpha]_D^{20}$ 为正值表示右旋（使偏振面顺时针偏转），负值表示左旋（使偏振面逆时针偏转）。由于果糖的左旋性大于葡萄糖的右旋性，随着水解反应的进行产物浓度增加，溶液系统的旋光度将由正值（右旋）经零度变为负值（左旋）。所以可用溶液旋光度的变化来度量反应进程。

设系统的起始旋光度为 α_0（$t=0$），反应终了时（$t=\infty$）的旋光度为 α_∞，反应到 t 时刻的旋光度为 α_t，则按式（2.41）有：

$$\alpha_0 = K_{反} c_{A,0} \tag{2.42a}$$

$$\alpha_\infty = K_{产} c_{A,0} \tag{2.42b}$$

式中，$K_{反}$ 和 $K_{产}$ 分别为反应物和产物在温度、样品管长度、光波长一定时的比例常数；$c_{A,0}$ 为蔗糖的初浓度亦即产物的最后浓度，当时间为 t 时，旋光度具有加和性，故

$$\alpha_t = K_{反} c_A + K_{产} (c_{A,0} - c_A) \tag{2.43}$$

联立式（2.42）和式（2.43）可得：

$$c_{A,0} = \frac{\alpha_0 - \alpha_\infty}{K_{反} - K_{产}} = K(\alpha_0 - \alpha_\infty) \tag{2.44}$$

$$c_A = \frac{\alpha_t - \alpha_\infty}{K_{反} - K_{产}} = K(\alpha_t - \alpha_\infty) \tag{2.45}$$

将上式代入式（2.39）即得：

$$\ln(\alpha_t - \alpha_\infty) = -kt + \ln(\alpha_0 - \alpha_\infty) \tag{2.46}$$

实测不同 t 时刻反应系统的 α_t 及反应完全时（$t=\infty$）的旋光度 α_∞，作 $\ln(\alpha_t - \alpha_\infty)$-$t$ 图为直线，由其斜率可求 k 值，进而计算 $t_{1/2}$。

三、仪器与试剂

旋光仪一套；超级恒温槽一台；恒温箱一只；秒表一块；50mL 容量瓶 1 个；200mL 锥形瓶 2 个；25mL 移液管 2 支；100mL 烧杯 1 个；蔗糖（AR）；3.0mol·L^{-1} 的 HCl 溶液。

四、实验步骤

1. 配制溶液

用台秤称取 10g 蔗糖于烧杯中，加少量蒸馏水溶解，然后倒入 50mL 容量瓶中稀释至刻度。用移液管各取 25mL 蔗糖和 HCl 溶液，分别置于两锥形瓶中，并将两锥形瓶浸于恒温槽（20℃）中恒温 10min。恒温槽的调节与使用见实验 2.1。

2. 测定旋光仪零点

旋光仪的原理及使用方法见第 4 章 4.9。蒸馏水为非旋光性物质，可用它找出仪器的零

点（α=0 时仪器所对应的刻度）。洗净旋光管，将一端加盖旋紧并盛满蒸馏水，使液体在一端形成一凸出的液面，然后从旁边推上玻璃片，以免管内存在气泡（有较大气泡时应重装），再旋上套盖直到不漏水为止。（注意旋紧套盖时勿用力过分，以免压碎玻璃片）。用毛巾或吸水纸将旋光管擦干。当两端玻璃片不干净时，要用擦镜纸擦拭。样品管中若有小气泡，应将其赶至样品管扩大部分。将旋光管放入旋光仪内，打开钠光灯，调整目镜焦距使视野清楚。旋转检偏镜手轮使在目镜中能观察到三分视野（零度视场，见第 4.9 节）为止，记下旋光仪上的读数 α（仅稍偏离 0），重复三次取其平均值，此值即为仪器的零点（是否为 $α_0$？），用来校正仪器的系统误差。

3. $α_t$ 的测定

待两锥形瓶溶液恒温后，将 HCl 溶液倒入蔗糖溶液之中（倒入一半时按下秒表作反应起点），迅速将混合液在两锥形瓶中反复倒两次，取少许溶液将旋光管洗 2～3 次，然后将溶液装入旋光管，盖好玻璃片并旋好套盖，擦净后立即放入旋光仪中，测定旋光度 $α_t$，按数据处理中表格给定时间读数记录。剩余的混合液加盖并置于 50～60℃ 恒温箱中加快反应速率，但温度不可超过 60℃，否则会产生副反应（颜色变黄）。

4. $α_∞$ 的测定

将恒温箱中反应完全后的溶液冷至实验温度，按步骤 2 装入旋光管中，测其旋光度即为 $α_∞$。

5. 实验完成

将旋光管、玻璃片、套盖内外一并洗净擦干，再装好，关上电源。由于溶液酸度大，易腐蚀仪器，使用时要注意防止溶液沾污仪器、皮肤及衣物等。

五、数据处理

1. 将实验测定结果列表 2.16 计算。

表 2.16　水解速率常数测定结果记录

实验温度：　　　℃　　　　　$α_∞$：

t/min	2	5	8	11	14	17	20	25	30	35	40	50	60
$α_t$													
$α_t-α_∞$													
$\ln(α_t-α_∞)$													

2. 作 $\ln(α_t-α_∞)$-t 图，由其斜率计算速率常数 k 和半衰期 $t_{1/2}$。

六、思考题

1. 本实验用蒸馏水校正旋光仪的零点，若不进行校正，对最后的实验结果有无影响？
2. 为什么配制蔗糖溶液可用台秤称量？
3. 混合蔗糖溶液和 HCl 溶液时，为何不把蔗糖溶液加入 HCl 溶液中去？
4. 已知旋光管长为 20 cm，试估算你所配制的蔗糖和 HCl 混合溶液的最初旋光度是多少？
5. 旋光仪的零点（零度视场）定为三分视野亮度一致者，在你所使用的仪器中有几个三分视野亮度一致的视场？为何选用亮度比较暗的那一个作为零度视场？

Experiment 2.9 Determination of Rate Constant for Hydrolysis of Sucrose

Objectives

To measure the reaction rate constant and half-life of the hydrolysis of sucrose. To understand the principle and master the usage of the polarimeter.

Principle

Under the catalysis of H^+ ion, the equation of hydrolysis of sucrose is:

$$C_{12}H_{22}O_{11} \text{ (sucrose)} + H_2O \xrightarrow{H^+} C_6H_{12}O_6 \text{ (dextrose)} + C_6H_{12}O_6 \text{ (fructose)}$$

The reaction is a second-order reaction, but since water is in such large excess that its concentration is almost no change, it is considered to be a pseudo first-order reaction. The rate equation is

$$-\frac{dc_A}{dt} = kc_A \tag{2.38}$$

The above equation is integrated, we obtain:

$$\ln c_A = -kt + \ln c_{A,0} \tag{2.39}$$

Where k is reaction-rate constant; $c_{A,0}$ is initial concentration of sucrose; c_A is concentration of sucrose at time t; t is reaction time.

The half-life of the reaction:

$$t_{1/2} = \frac{\ln 2}{k} \tag{2.40}$$

Measure the concentrations of sucrose at different time and plot $\ln c_A$ against time. Then from the slope of the straight line we can calculate the reaction-rate constant k.

In this experiment, the rate of reaction is followed by observing the change of a physical property, the optical rotation.

Each substance has its own optical rotation:

sucrose $[\alpha]_D^{20} = 66.65°$ (dextrorotatory)
glucose $[\alpha]_D^{20} = 52.5°$ (dextrorotatory)
fructose $[\alpha]_D^{20} = -90.9°$ (levorotatory)

Sucrose is dextrorotatory, but the resulting mixture of glucose and fructose is slightly levorotatory because the levorotatory fructose has a greater molar rotation than the dextrorotatory glucose. As the sucrose is used up and the glucose-fructose mixture is formed, the angle of rotation to the right becomes less and less, and finally the light is rotated to the left.

It is assumed that α_0 stands for optical rotation at the beginning of the reaction, α_t stands for optical rotartion at time t, and α_∞ stands for optical activity in the end.

Given by

$$\alpha = [\alpha]_\lambda^t \frac{lc}{100} \tag{2.41}$$

We can obtain:

$$\alpha_0 = K_{反} c_{A,0} \tag{2.42a}$$

$$\alpha_\infty = K_{产} c_{A,0} \tag{2.42b}$$

$$\alpha_t = K_{反} c_A + K_{产}(c_{A,0} - c_A) \tag{2.43}$$

Combining Eqs. (2.42) and (2.43), we obtain:

$$c_{A,0} = \frac{\alpha_0 - \alpha_\infty}{K_{反} - K_{产}} = K(\alpha_0 - \alpha_\infty) \tag{2.44}$$

$$c_A = \frac{\alpha_t - \alpha_\infty}{K_{反} - K_{产}} = K(\alpha_t - \alpha_\infty) \tag{2.45}$$

Adding Eqs. (2.44) and (2.45) to Eqs. (2.39):

$$\ln(\alpha_t - \alpha_\infty) = -kt + \ln(\alpha_0 - \alpha_\infty) \tag{2.46}$$

In the experiment, measure the optical rotartion at time t (α_t) and at the end of the reaction (α_∞) respectively. Plot the $\ln(\alpha_t - \alpha_\infty)$ against time, from which we can figure out the reaction-rate constant k that is negative to the slope. Then we can calculate the half-life from the constant k.

Apparatus and Reagents

Polarimeter	1
Super thermostatic trough	1
Thermostat container	1
Stopwatch	1
Volumetric flask (50mL)	1
Conical flask (200mL)	2
Pipette (25mL)	2
Beaker (100mL)	1

Sucrose (analytical pure)

HCl solution ($c = 3.0$ mol/L)

Procedure

1. Make up the solution

Dissolve 10g of sucrose and make up the solution in a 50mL of volumetric flask. Pipette 25mL solution into a conical flask. Pipette 25mL of HCl solution into another conical flask. Then take them into thermostatic trough. Adjust temperature of the thermostatic bath at 20℃ ±0.1℃.

2. Regulate the zero point

Electrify the polarimetry. Fill the optical pile full of distilled water, without any air bladder and leak, then set it into the polarimeter. Regulate the focus to make the trisection visual field clear. Rotate the microscopy, when trisection visual field disappears, mark down the angle that is just the zero point. Repeat this procedure three times. Subsequently, calculate the average value, then pour out water and get ready for next operation.

3. Determination of α_t

After temperature is kept constant for 10min, add HCl solution into sucrose and record the time with a stopwatch when adding one half. Wash the optical pile with mixture for 2~3 times, fill it with mixture, without any air bladder and leak, then put it into polarimeter again. Rotate the microscopy, mark down the angle as soon as trisection visual field disappear. Record each time and reading according to the following table 2.17. Heat the mixture remained in a water bath at 50~60℃.

4. Determination of α_∞

After the reaction is complete, cool the mixture down to experiment temperature. Repeat the procedure 2, 3 and determine α_∞.

5. Experiment accomplished

After the experiment is finished, the apparatus must be washed cleanly and wiped dry.

Observations and Measurements

Table 2.17　Data record of the experiment

experiment temperature：　　　　　　　　　　　　α_∞：

t/min	2	5	8	11	14	17	20	25	30	35	40	50	60
α_t													
$\alpha_t - \alpha_\infty$													
$\ln(\alpha_t - \alpha_\infty)$													

Calculations

Draw a graph of $\ln(\alpha_t - \alpha_\infty)$ against time. Calculate the reaction-rate constant k and half-life from the slope.

Notes

1. Optical pipe must be filled full of liquid without any air bladder.

2. Each time the polarimeter tube is used, rinse it with a portion of the reaction mixture.

3. For the mixture is rather acid and corrosive, the apparatus must be washed clean and wiped dry after the experiment is finished.

Questions

1. If do not regulate the zero point of polarimetry, will the result of experiment be influenced?

2. Why sucrose can be weighted by table balance?

3. Why cannot add sucrose into HCl solution when mixing?

4. Why it is zero point when the brightness of the trisection visual field is identical and darker?

Appendix

1. Switch on the electrical source (220V) and turn on the Na-light. As the lamp-house is stable, regulate the focus to make the trisection visual field clear showed as ⬤ or ⬤.

2. Regulate the zero point.

Fill the optical pipe full of distilled water. When rotate the microscopy, mark down the point on which trisection visual field disappears. It is just like the picture 2.

3. Repeat the same procedure and determine the optical activity of sample.

After the experiment is finished, the apparatus must be cleaned and wiped dry.

实验 2.10 过氧化氢分解速率常数的测定

一、 实验目的

测定过氧化氢被碘化钾催化分解的速率常数，了解一级反应的基本特征，掌握量体积法（物理法）测量动力学参数的方法。

二、 实验原理

过氧化氢分解的化学计量式为：

$$H_2O_2 \longrightarrow H_2O + \frac{1}{2}O_2 \tag{2.47}$$

该反应有催化剂（如 KI）存在时，可以使其分解速率加快。H_2O_2 在 KI 作用下催化分解按下列步骤进行：

$$H_2O_2 + KI \longrightarrow KIO + H_2O \tag{2.48}$$

$$KIO \longrightarrow \frac{1}{2}O_2 + KI \tag{2.49}$$

由上式可看出，KI 与 H_2O_2 生成中间化合物，改变了反应途径，降低了反应活化能而使反应速率加快。据分析，反应式（2.49）的速率远较反应式（2.48）的速率快，故反应式（2.48）成为整个反应的控制步骤。因此，总反应速率等于式（2.48）的速率，即

$$-\frac{dc_{H_2O_2}}{dt} = k' c_{KI} c_{H_2O_2} \tag{2.50}$$

由于 KI 浓度不变，故与 k' 合并仍为常数，令 $k' c_{KI} = k$，故方程式（2.50）可简化为

$$-\frac{dc_{H_2O_2}}{dt} = k c_{H_2O_2} \tag{2.51}$$

由式（2.51）看出，反应速率与反应物浓度的一次方成正比。故知 H_2O_2 在 KI 作用下的催化分解反应是一级反应，且反应速率常数 k 将随 KI 的浓度变化而改变。

将式（2.51）积分，得

$$\ln \frac{c_t}{c_0} = -kt \tag{2.52}$$

式中，c_0 为 H_2O_2 的初始浓度；c_t 为 t 时刻 H_2O_2 的浓度。

如以 $\ln c_t$ 对 t 作图得一直线，即可确定 H_2O_2 催化分解反应为一级反应。由直线斜率便可求出反应速率常数 k。

怎样求得反应过程中 H_2O_2 的浓度 c_t 呢？本实验采用物理法进行测定。从化学计量式

(2.47) 看出，1 mol H_2O_2 分解可得 0.5 mol O_2。故可通过测量某一时刻 t 放出的氧气体积即可得到相应时刻的 H_2O_2 浓度 c_t。其关系如下：

令 V_∞ 表示 H_2O_2 全部分解放出的 O_2 体积；V_t 表示 H_2O_2 在 t 时刻分解放出的 O_2 体积。

则 $c_0 \propto V_\infty$ 或 $c_0 = fV_\infty$

$c_t \propto V_\infty - V_t$ 或 $c_t = f(V_\infty - V_t)$

将以上关系式代入式 (2.52)，经整理便得下式：

$$\ln(V_\infty - V_t) = -kt + \ln V_\infty \tag{2.53}$$

以 $\ln(V_\infty - V_t)$ 对 t 作图得一直线，则可验证 H_2O_2 的催化分解反应为一级。由直线斜率可求得反应速率常数 k，进而求得反应的半衰期 $t_{1/2}$：

$$t_{1/2} = \ln 2/k \tag{2.54}$$

V_∞ 的确定有如下三种方法。

（1）V_∞ 可由 H_2O_2 溶液的起始浓度及体积求出。H_2O_2 溶液的起始浓度是用高锰酸钾标准溶液在酸性介质中进行滴定求得，其化学计量式如下：

$$5H_2O_2 + 2KMnO_4 + 3H_2SO_4 = 2MnSO_4 + K_2SO_4 + 8H_2O + 5O_2$$

由滴定所用 $KMnO_4$ 标准溶液的体积和浓度便可算出 H_2O_2 的准确浓度，由 H_2O_2 的分解反应可知，1 mol H_2O_2 可分解出 0.5 mol O_2，于是可求出 H_2O_2 完全分解时 O_2 的体积 V_∞。

（2）将反应瓶在水浴中加热至 50~60℃，并保持 15min，再冷却至实验温度，读取量气管体积即为 V_∞。

（3）由实测不同 t 时刻氧气的体积 V_t，作 V_t-$1/t$ 图，外推至 $1/t \to 0$ 即为 V_∞。

本实验采用方法（2）即加热法来确定 V_∞。

三、仪器与试剂

实验装置如图 2.31 所示；超级恒温槽一台；反应瓶 1 个；气体体积测量装置一套；塑料瓶盖一个；5mL、10mL 移液管各 1 支；20mL 量筒 1 个；秒表 1 块；0.1mol·L^{-1} KI 溶液；2% H_2O_2 溶液。

四、实验步骤

1. 接通恒温水浴，调节恒温水温度为 25℃（恒温槽的调节与使用参见实验 2.1 或第 4.2 节）。

2. 用量筒量取 20mL 蒸馏水，用移液管量取 10mL 0.1mol·L^{-1} KI 溶液，置于反应瓶中；用移液管量取 2% H_2O_2 溶液 5mL，小心注入反应瓶中的塑料瓶盖（或乒乓球半球）中，注意勿使 H_2O_2 与 KI 溶液相混，盖紧反应瓶上的瓶塞。

3. 旋转三通活塞于（a）位（见图 2.31），高举水准瓶，让液体充满量气管，之后旋转三通到（b）位并把水准瓶放低。如量气管内液面在 2min 内保持不变，即表示系统不漏气，否则应找出原因，排除后再试漏。读取量气管初始体积 V_0（读数时，一定要使水准瓶内液面和量气管内液面保持同一水平）。

4. 倾斜反应瓶，使 H_2O_2 溶液与 KI 溶液相混，及时按下秒表计时，并均匀地摇动反应瓶，开动搅拌器搅拌。

5. 每隔 2min 读取量气管读数一次，共读 7 次，然后每隔 5min 读一次，直至 40min。

6. 用加热的方法测 V_∞。

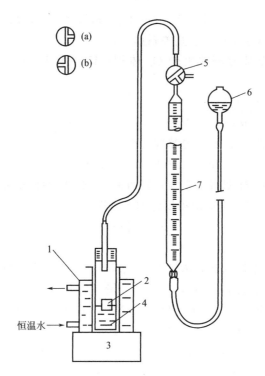

图 2.31 过氧化氢分解实验装置示意图
1—带夹套的反应瓶；2—塑料瓶盖；3—电磁搅拌器；
4—搅拌子；5—三通旋塞；6—水准瓶；7—量气管

五、数据处理

1. 用 V_0 校正所有的体积读数 V_t，V_∞。
2. 列出 t，V_t，$(V_\infty - V_t)$，$\ln(V_\infty - V_t)$ 数据（表 2.18）。
3. 以 $\ln(V_\infty - V_t)$ 对 t 作图，求直线斜率。
4. 计算 H_2O_2 分解反应速率常数 k 和 $t_{1/2}$。

表 2.18 分解速率常数测定结果记录

t/min	2	4	6	8	10	12	14	19	24	29	34	39
V_t												
$V_\infty - V_t$												
$\ln(V_\infty - V_t)$												

六、注意事项

1. 反应前 H_2O_2 与 KI 不要接触。
2. 不要拧错三通的位置。
3. 若系统漏气，多为反应瓶上的橡皮塞未盖紧。
4. 每次读数时务必使水准瓶内液面和量气管内液面处于同一水平面。

七、思考题

1. H_2O_2 催化分解是几级反应？其特征是什么？试用所得结果进行分析。
2. 反应速率常数与哪些因素有关？

3. 本实验为什么是以 $\ln(V_\infty - V_t)$ 对 t 作图而不直接以 $\ln\frac{c_t}{c_0}$ 对 t 作图呢？

4. 本实验装置中，夹套瓶及搅拌器各起什么作用？搅拌速度对结果有无影响？

5. 你对改进本实验方法有何建议？

Experiment 2.10 Catalytic Decomposition of Hydrogen Peroxide

Objectives

To measure the reaction rate constant and half-life of the iodide-catalyzed decomposition of H_2O_2. To learn properties of the first-order reaction.

Principle

The reaction of decomposition of H_2O_2 is

$$H_2O_2 \longrightarrow H_2O + \frac{1}{2}O_2 \tag{2.48}$$

The decomposition, normally slow, may be catalyzed by iodide ion. With the catalysis of KI, hydrogen peroxide decomposes in accordance with the following equations:

$$H_2O_2 + KI \longrightarrow KIO + H_2O \quad (\text{slow}) \tag{2.49}$$

$$KIO \longrightarrow \frac{1}{2}O_2 + KI \quad (\text{quick}) \tag{2.50}$$

The reaction (2.49) is the rate controlling step. So the rate equation is

$$-\frac{dc_{H_2O_2}}{dt} = k' c_{KI} c_{H_2O_2} \tag{2.51}$$

In the experiment, the concentration of KI is constant. $k' c_{KI} = k$. The above equation is written as:

$$-\frac{dc_{H_2O_2}}{dt} = k c_{H_2O_2} \tag{2.52}$$

The reaction is a first-order reaction.

The above equation is integrated, we obtain:

$$\ln\frac{c_t}{c_0} = -kt \tag{2.53}$$

Where k is reaction-rate constant; c_0 is initial concentration of H_2O_2; c_t is concentration of H_2O_2 at time t; t is reaction time.

Measure concentrations of H_2O_2 at time in minutes and plot $\ln c_t$ against time. From the slope of the straight line we can calculate the reaction-rate constant k.

In this experiment, the progress of a reaction is studied by measuring the quantity of gas produced as a function of time. At any given time, the volume of oxygen evolved is directly proportional to the number of moles of hydrogen peroxide that have decomposed, provided, that the volume of oxygen is measured at constant pressure and temperature.

If V_∞ represents the final gas volume and V_t represents the gas volume at time t, We can obtain:

$$c_0 \propto V_\infty \text{ 或 } c_0 = fV_\infty$$
$$c_t \propto V_\infty - V_t \text{ 或 } c_t = f(V_\infty - V_t)$$

Adding c_0, c_t to Eq. (2.53):
$$\ln(V_\infty - V_t) = -kt + \ln V_\infty$$

Plot $\ln(V_\infty - V_t)$ against time. From the slope of the straight line we can figure out the reaction-rate constant k. Then we can calculate the half-life.

$$t_{1/2} = \frac{\ln 2}{k} \tag{2.54}$$

Determine V_∞ by the method of heating. Adjust temperature of the thermostatic bath at 50~60 ℃. Keep the reaction flask at constant temperature for 15min. Cool the flask down to experimental temperature before reading the final volume.

Apparatus and Chemicals Required

Gas volume measuring device	(100 mL gas buret) (Fig. 2.32)	1
Super thermostatic trough		1
Stopwatch		1
Reaction flask		1
Pipette (5 mL)		1
Pipette (10 mL)		1
Graduated cylinder (20 mL)		1

KI solution   0.1 mol·L^{-1}
H_2O_2 solution   2%

Procedure

1. Adjust temperature of the thermostatic bath at 25 ℃±0.1 ℃.

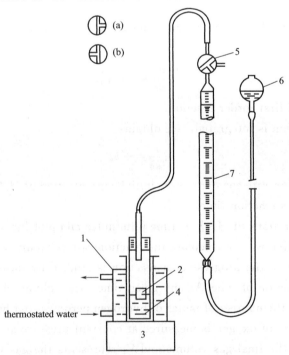

Fig. 2.32   The equipments of decomposition of H_2O_2

2. Graduate 20 mL of distilled water and 10.00 mL of KI solution with pipette into a reaction flask. Transfer 5.00 mL of H_2O_2 solution with pipette a plastic cover placed in the reaction flask. Cover the piston tightly.

3. Revolve three-way cock to (a), the leveling bulb is raised so that the gas buret is brimming. Revolve three-way cock to (b), and the leveling bulb is lowered. If the liquid level in the buret do not fall down in 2 min, the system has good airtightness, or else must find out the problem before next procedure. The leveling bulb is raised or lowered until the liquid is at the same level in both the buret and the bulb. At that moment, the initial gas volume V_0 is read.

4. Incline the reaction flask, the reactants are mixed meanwhile record the time with a stopwatch. Shake the flask at a slow regular rate.

5. The buret is read per 2 min interval for seven times, then read per 5 min interval until 40 min.

6. Determine V_∞ by the method of heating. Adjust temperature of the thermostatic bath at 50~60 ℃. Keep the reaction flask constant temperature for 15 min. Cool the flask down to 25 ℃ before reading the final volume.

Observations and Measurements

Table 2.19　Data record of the experiment

experiment temperature:
atmospheric pressure:
V_∞:
the value of V_0 was corrected with V_t, V_∞.

t/min	2	4	6	8	10	12	14	19	24	29	34	39
V_t												
$V_\infty - V_t$												
$\ln(V_\infty - V_t)$												

Calculations

Draw a graph of $\ln(V_\infty - V_t)$ against time. Calculate the reaction-rate constant k and half-life from the slope.

Notes

1. Do not mix the reactants before recording the time.
2. Cover the piston tightly.
3. Read volumes when the liquid is at the same level in both the buret and the bulb.

Questions

1. What is the characteristic of the iodide-catalyzed decomposition of H_2O_2 according to the results?

2. What are factors affecting the reaction-rate constant k?

3. Why not draw a graph of $\ln \dfrac{c_t}{c_0}$ against time?

4. What are the function of double-jacket flask in the experiment?

5. What are your suggestions about the experiment?

实验 2.11 乙酸乙酯皂化反应动力学参数的测定

一、实验目的

掌握由电导率仪测定乙酸乙酯皂化反应动力学参数的方法；了解二级反应动力学规律及特征。掌握 DDS-11A 型电导率仪的使用方法；求不同温度下皂化反应的 k、$t_{1/2}$ 和 E_a。

二、实验原理

乙酸乙酯皂化反应为双分子反应，其反应式为：

$$CH_3COOCH_2CH_3 + OH^- \rightleftharpoons CH_3COO^- + C_2H_5OH$$

对于二级反应，若两反应物初始浓度相同，均为 a，则可得：

$$\frac{dx}{dt} = k(a-x)^2 \tag{2.55}$$

x 是经过 t 时间后减小了的反应物浓度，上式积分得：

$$k = \frac{1}{t} \times \frac{x}{a(a-x)} \text{ 或 } \frac{x}{a(a-x)} = kt \tag{2.56}$$

起始浓度 a 为已知，因此只要由实验测得不同时间 t 时的 x 值，以 $x/(a-x)$ 对 t 作图，若所得为一直线，证明是二级反应，并可以从直线的斜率求出 k 值。

不同 t 时的 x 值可用化学分析法测定，也可用物理法（如电导法）测定。这里采用电导法测量 x 的依据是：

① 在稀溶液中，强电解质的电导率与其浓度成正比，而且溶液总电导率等于组成溶液的各电解质电导率之和。

② 本实验中反应物只有 NaOH 为强电解质，生成物只有 CH_3COONa 是强电解质，且 OH^- 的电导率比 CH_3COO^- 大得多。随反应进行，OH^- 浓度不断减少，溶液电导率也随之下降。

因此，可用电导率仪测量皂化反应进程中电导率随时间的变化，从而达到跟踪反应物浓度随时间变化的目的。

可以假定，稀溶液中溶液电导率的降低与 OH^- 浓度的减小成正比，即：

$$\kappa = \alpha + \beta c \tag{2.57}$$

式中，κ 为溶液电导率；c 为 OH^- 浓度；α、β 为常数。

当 $t=0$ 时，$\kappa_0 = \alpha + \beta c_0$

$t=t$ 时，$\kappa_t = \alpha + \beta c_t$

$t=\infty$ 时，$\kappa_\infty = \alpha$

所以

$$\frac{x}{a-x} = \frac{c_0 - c_t}{c_t} = \frac{\kappa_0 - \kappa_t}{\kappa_t - \alpha} = \frac{\kappa_0 - \kappa_t}{\kappa_t - \kappa_\infty} \tag{2.58}$$

代入式（2.56）可得

$$k = \frac{1}{ta} \frac{\kappa_0 - \kappa_t}{\kappa_t - \kappa_\infty}$$

或

$$\kappa_t = \frac{1}{ka} \frac{\kappa_0 - \kappa_t}{t} + \kappa_\infty \tag{2.59}$$

通过实验测定不同时间溶液的电导率 κ_t 和起始溶液的电导率 κ_0，以 κ_t 对 $(\kappa_0 - \kappa_t)/t$ 作

图，也得一直线，从直线的斜率也可求出反应速率常数 k 值。

按下式可求出反应的半衰期：
$$t_{1/2}=1/(ka) \quad (2.60)$$

测量不同温度 T_1 和 T_2 的速率常数 $k(T_1)$ 和 $k(T_2)$，由阿伦尼乌斯公式可求得反应的活化能 E_a：

$$\ln\frac{k(T_2)}{k(T_1)}=\frac{E_a}{R}\left(\frac{1}{T_1}-\frac{1}{T_2}\right) \quad (2.61)$$

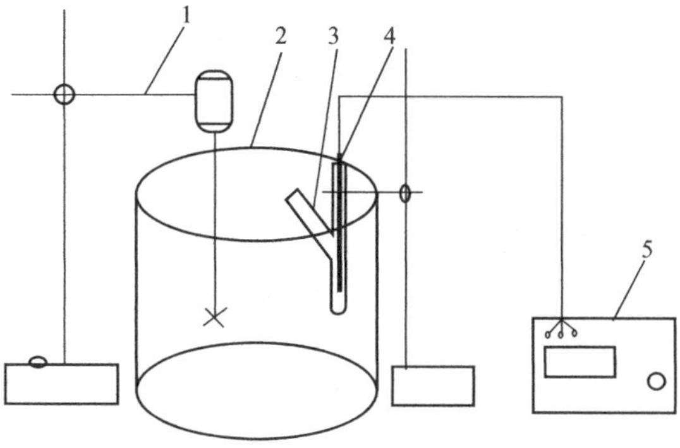

图 2.33　皂化反应动力学测量装置示意图
1—搅拌器；2—恒温槽；3—羊角池；
4—电极；5—电导率仪

三、仪器与试剂

DDS-11A 型电导率仪；光亮电极；恒温槽；羊角形电导池 2 支；大试管 2 支；5mL 移液管 2 支；$0.02\text{mol}\cdot\text{L}^{-1}$ NaOH 标准溶液；$0.02\text{mol}\cdot\text{L}^{-1}$ $CH_3COOC_2H_5$ 标准溶液。实验装置见图 2.33。

四、实验步骤

1. 调节恒温槽（参见实验 2.1）至 25℃（夏天可调至 30℃）。
2. 调节电导率仪（参见第 4 章 4.7）。
3. 测 κ_0：在洗净烘干后的大试管内用移液管移入蒸馏水和 NaOH 溶液各 5mL，插入电导电极，恒温后用电导率仪测 κ_0 值（每次读数前应校正，下同）。
4. 测 κ_t：用移液管吸取 NaOH 和 $CH_3COOC_2H_5$ 溶液各 5mL，分别小心放入洗净烘干的羊角形电导池的两支管（切勿使两液相混），将羊角形电导池置于恒温槽中，恒温后迅速将两种液体混合（混合一半时按下秒表记录反应时间），来回倒两次，插入电导电极（用蒸馏水洗涤并用滤纸吸干），分别在第 2、4、6、9、12、15、20、25、30、40min 时测其电导率值 κ_t。
5. 调节恒温槽至 35℃，重复步骤 3 和 4 测定之。

五、数据处理

1. 将不同温度下测定的数据列入表 2.20。

表 2.20　反应动力学参数测定实验记录

温度：　　℃　　　初浓度 a：　　　电导率仪编号：

t/min	2	4	6	9	12	15	20	25	30	40
κ_t										
$\kappa_0-\kappa_t$										
$(\kappa_0-\kappa_t)/t$										

2. 以 κ_t 对 $(\kappa_0-\kappa_t)/t$ 作图，由直线斜率求 25℃ 和 35℃ 的速率常数 k 和半衰期 $t_{1/2}$（注意溶液的初浓度 a 应作换算）。
3. 由阿伦尼乌斯公式计算反应的活化能 E_a。

六、注意事项

1. 大试管、羊角形电导池必须洗净烘干。

2. 计时前羊角形电导池中的两种溶液切勿相混。

3. 每次读数前都应校正电导率仪。

4. 每次测定后电导电极应用蒸馏水洗涤，并浸泡在干净的蒸馏水中，每次测定前用滤纸小心吸干电极上的蒸馏水。

七、思考题

1. 乙酸乙酯皂化反应的动力学特征是什么？
2. 为什么要使两种反应物的初浓度相等？
3. 本实验测定的误差主要来自哪几方面？你对改进本实验方法有何建议？

Experiment 2.11　Determination of Kinetic Parameters for Saponification of Ethyl Acetate

Objectives

To master the conductivity method to determine acetidin saponification rate constant, half-life and activation energy. To learn properties of the second-order reaction. To master the operation of DDS-11A conductivity detector.

Principle

The reaction studied in this experiment is

$$CH_3COOCH_2CH_3 + OH^- \rightleftharpoons CH_3COO^- + C_2H_5OH$$

It is a second-order reaction. For the case initial concentrations of the reacting substances is equal, the reaction follows the equation

$$\frac{dx}{dt} = k(a-x)^2 \tag{2.55}$$

The above equation is integrated, we obtain:

$$k = \frac{1}{t} \times \frac{x}{a(a-x)} \text{ 或 } \frac{x}{a(a-x)} = kt \tag{2.56}$$

Where k is reaction-rate constant; a is initial concentrations of the reacting substances; x is concentration reacted at time t; t is reaction time.

If the value of "a" is known, measure x at different time and plot $x/(a-x)$ against time. A straight line is obtained. It is proved that order of the reaction is second. From the slope we can calculate the reaction-rate constant K.

Measure x at time t by the conductivity method, because:

(1) The conductivity of strong electrolyte solution is directly proportional to its concentration in dilute solution.

(2) In this experiment, NaOH and CH_3COONa are strong electrolytes. Na^+、CH_3COO^-、OH^- ions are electric in the solution. Concentration of Na^+ has no change, which do not contribute to the changes in conductivity. Because molar conductivity of OH^- is much bigger than CH_3COO^-, the conductivity of the system can be considered to decline when

concentration of OH^- decreases and CH_3COO^- increases through the reaction progress.

The progress of the reaction can therefore be followed by measurement of conductivity.

It is assumed that decline of the conductivity of the system is directly proportional to decrease of the concentration of OH^- in dilute solution.

$$\kappa = \alpha + \beta c \tag{2.57}$$

Where κ is conductivity in the reaction; c is concentration of NaOH; α、β are constants. We can obtain:

At the beginning ($t=0$), $\kappa_0 = \alpha + \beta c_0$

At time t ($t=t$), $\kappa_t = \alpha + \beta c_t$

In the end ($t=\infty$), $\kappa_\infty = \alpha$

we obtain:

$$\frac{x}{a-x} = \frac{c_0 - c_t}{c_t} = \frac{\kappa_0 - \kappa_t}{\kappa_t - \alpha} = \frac{\kappa_0 - \kappa_t}{\kappa_t - \kappa_\infty} \tag{2.58}$$

Adding Equation (2.58) to Equation (2.56), we obtain

$$k = \frac{1}{ta} \frac{\kappa_0 - \kappa_t}{\kappa_t - \kappa_\infty}$$

which upon rearrangement becomes

$$\kappa_t = \frac{1}{ka} \frac{\kappa_0 - \kappa_t}{t} + \kappa_\infty \tag{2.59}$$

In the experiment, measure conductivity at time t (κ_t) and conductivity at the beginning of the reaction (κ_0) and plot κ_t against ($\kappa_0 - \kappa_t$)$/t$. From the slope of the straight line we can figure out the reaction-rate constant k. Then we can calculate the half-life.

$$t_{1/2} = 1/(ka) \tag{2.60}$$

Measure the values of k at two temperatures and calculate activation energy of the reaction.

$$\ln \frac{k(T_2)}{k(T_1)} = \frac{E_a}{R}\left(\frac{1}{T_1} - \frac{1}{T_2}\right) \tag{2.61}$$

Apparatus and Chemicals Required

DDS-ⅡA Conductivity detector	1
Platinum black electrode	1
Thermostatic trough	1
Fork-shaped conductance cell	2
Big test tube	2
Pipette (5mL)	2
Stopping watch	1

NaOH $0.02\,mol \cdot L^{-1}$

$CH_3COOC_2H_5$ $0.02\,mol \cdot L^{-1}$

Procedure

1. Adjust the temperature of thermostatic trough at 25℃.

2. Regulate the conductivity detector. Command the regulating technique.

3. Determine κ_0. Pipette 5mL of NaOH solution and 5mL of distilled water into a big test tube which is rinsed and dry. Put a electrode into the solution, then take the big test

tube into thermostatic trough. Determine κ_0 by the conductivity detector after solution being kept temperature constant. (Conductivity detector must be proof-read before each determination.)

4. Determine κ_t. Pipette 5mL of NaOH solution into one branched-tube of fork-shaped conductance cell which is rinsed and dry and pipette 5mL of $CH_3COOC_2H_5$ solution into another branched-tube (Do not mix the reactants). Put the electrode into the solution. Place the conductance cell in the thermostatic trough. After temperature is kept constant, the reactants are mixed rapidly. Record the time with a stopwatch when mixed one half. Determine κ_t at difference time of 4, 6, 9, 12, 15, 20, 25, 30 and 40min.

5. Adjust the temperature of thermostatic trough at 35℃. Repeat procedure 3, 4。

Observations and Measurements

Table 2.21 Data record of the experiment

room temperature:

atmospheric pressure:

the initial concentration of NaOH:

serial number of conductivity detector:

t/min		4	6	9	12	15	20	25	30	40
κ_t										
$\kappa_0 - \kappa_t$										
$(\kappa_0 - \kappa_t)/t$										

Calculations

1. Plot κ_t against $(\kappa_0 - \kappa_t)/t$. From the slope calculate the reaction-rate constant k and half-life at 25℃ and 35℃.

2. Calculate activation energy of the reaction.

Notes

1. Big test tubes and fork-shaped conductance cells must be rinsed and dry.

2. Do not mix the reactants before recording the time.

3. Conductivity detector must be proof-read before each determination.

4. Clean a platinum black electrode with distilled water and blotted it up by the use of filter paper (Don't touch platinum-black).

Questions

1. What is the characteristic of the acetidin saponification?

2. Why the initial concentrations of the reactants are equal?

3. What are the experimental errors? What are your suggestions about the experiment?

Appendix

1. Switch on the electrical source and allow it warm-up for several minutes (until pointer turns stable).

2. Turn the constant knob to be electrode constant (It's been checked and determined by the factory).

3. Determine conductivity of NaOH on high round.

4. Determination. Choose proper measuring range from large range to small one and try to control the pointer pointing to the full scale in order to get accurate data.

That just is: $10^4 \rightarrow 10^3 \rightarrow 10^2 \rightarrow 10 \rightarrow 1 \rightarrow 0.1$

5. Reading and record. The data marker is in the form of $\times 10x \times 10^{-6}$ (s·cm^{-1}). X stands for 4, 3, 2 and 1 in the measuring range.

实验 2.12 溶液表面张力的测定

一、实验目的

1. 测定不同浓度正丁醇水溶液的表面张力。
2. 求正丁醇在界面上的吸附量和正丁醇分子的横截面积。
3. 掌握最大气泡法测定表面张力的原理和技术。

二、实验原理

在各种不同相的界面上都可以发生吸附现象。溶液表面也可产生吸附作用，当在液体中加入某种溶质时，液体的表面张力会发生变化。若溶质使液体表面张力升高，则溶质在溶液表面层的浓度小于在溶液内部的浓度；若溶质使液体表面张力降低，则溶质在溶液表面层的浓度大于其在溶液内部的浓度。这种溶质在溶液表面的浓度与溶液内部的浓度不同的现象称为溶质的表面吸附。

在一定的温度、压力下，溶质的表面吸附量与溶液的浓度、溶液的表面张力间的关系，可以用 Gibbs 吸附等温式表示：

$$\Gamma = -\frac{c}{RT}\frac{d\sigma}{dc} \tag{2.62}$$

式中，Γ 为表面过剩吸附量，mol·m^{-2}；c 为溶液浓度；σ 为溶液表面张力，N·m^{-1}；R 为通用气体常数；T 是热力学温度，K。若 $\frac{d\sigma}{dc}<0$，则 $\Gamma>0$，即随着溶液浓度增加，溶液表面张力降低，表面吸附量为正，称为正吸附；若 $\frac{d\sigma}{dc}>0$，则 $\Gamma<0$，即随着溶液浓度增加，溶液表面的张力增加，表面吸附量为负，称为负吸附。

使溶液表面张力降低的物质，称为表面活性物质。这类物质的分子结构一般是由极性基团和非极性基团构成。在水溶液的表面，表面活性剂分子的极性部分指向溶液内部，非极性部分则指向空气。对于正丁醇来说，—OH 为极性部分而—RCH$_2$ 则为非极性部分。当溶液的浓度增大到某一程度时，被吸附的表面活性物质分子占满了所有表面，形成了单分子的饱和吸附层。

图 2.34 表示正丁醇水溶液的 σ-c 曲线。从 σ-c 曲线上可求得不同浓度时的 $\left(\frac{d\sigma}{dc}\right)$ 值，将它带入 Gibbs 公式即可算出不同浓度时的吸附量 Γ。作 Γ-c 曲线见图 2.35，即可求出不同浓度时的吸附量 Γ_∞，设在饱和吸附下，正丁醇分子在气-液界面上辅满一单分子层，应用下式求得正丁醇分子的横截面积 S：

$$S = \frac{1}{\Gamma_\infty N_0} \tag{2.63}$$

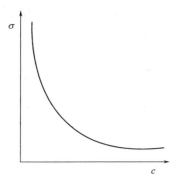

图 2.34 σ-c 曲线

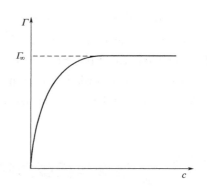

图 2.35 溶液吸附等温线

式中，N_0 为阿伏加德罗常数。

实际上，从 Γ-c 曲线外推求 Γ_∞ 比较困难。若设适用于气-固单分子层吸附的 Langmuir 吸附等温式，对气-液单分子层同样适用，则应有：

$$\frac{c}{\Gamma}=\frac{c}{\Gamma_\infty}+\frac{1}{K\Gamma_\infty} \tag{2.64}$$

式中，c 为溶液的平衡浓度；Γ_∞ 为饱和吸附量；K 为常数，与溶质的表面活性大小有关。作 c/Γ-c 图为直线，由斜率可求 Γ_∞，然后由式（2.63）求正丁醇分子截面积 S。

本实验通过测定正丁醇水溶液的 σ-c 曲线，从而算出不同浓度下的 $\dfrac{\mathrm{d}\sigma}{\mathrm{d}c}$ 值，然后由 c/Γ-c 直线的斜率得出 Γ_∞ 以求取正丁醇分子的横截面积 S。

最大气泡法测定表面张力的装置见图 2.36，其原理介绍如下。

将毛细管的端面液面相切，液面即沿毛细管上升，打开抽气滴液漏斗的活塞，让水缓缓滴下，使毛细管内的溶液受到的压力比样品中试样液面上来得大。当此压力差于毛细管端面上产生的作用力稍大于毛细管口液体的表面张力时，毛细管口的气泡即被压出，压差的最大值 $p_{\max}$ 可从压力计上读出。

$$p_{\max}=p_{大气}-p_{系统}=\Delta h\rho g$$

式中，Δh 是 U 形压力计之压差值；ρ 为压力计内介质的密度；g 为重力加速度。

若毛细管的内径为 r，则这个最大压力差产生的驱使气泡逸出液面的作用力应为

$$F=\pi r^2 p_{\max}=\pi r^2 \Delta h\rho g$$

气泡在毛细管口受到表面张力引起的作用力应为

$$F=2\pi r\sigma$$

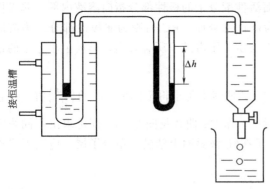

图 2.36 最大气泡法装置图

在气泡刚离开管口时，上述两作用力相等，即

$$\pi r^2 \Delta h\rho g=2\pi r\sigma$$

$$\sigma=\frac{1}{2}r\Delta h\rho g \tag{2.65}$$

在实验中，若使用同一支毛细管和压力计，则 $\dfrac{1}{2}r\rho g$ 是一个常数，称作仪器常数，用 K 来表示：

$$\sigma=K\Delta h \tag{2.66}$$

如果将已知表面张力的溶液作为标准，

由实验测定得其 Δh 后,就可求出 K 值,然后只要用这一仪器测定其他溶液的 Δh 值,即可求得各种溶液的表面张力 σ。

三、仪器与试剂

表面张力测定装置 1 套;100mL 容量瓶 9 个;正丁醇(AR)。

四、实验步骤

1. 各种浓度正丁醇水溶液的配制

将 10mL 比重瓶一个洗净烘干,冷却后称重,然后用干净吸管注满正丁醇后,迅速于分析天平上称重。若得 10mL 液体的质量为 W(g),可按下式计算其密度 ρ(g·mL^{-1}):

$$\rho = W/10$$

浓度为 0.010、0.020、0.050、0.100、0.150、0.200、0.250、0.300、0.350mol·L^{-1} 的正丁醇水溶液可按下法配制:

用刻度移液管吸取正丁醇 $\dfrac{0.100 \times 74}{10 \times \rho}$ mL 于容量瓶中,用蒸馏水稀释至刻度,则得其水溶液浓度为 0.010 mol·L^{-1}。其他浓度的溶液可用同法计算配制。

2. 溶液表面张力的测定

洗净表面张力仪的各部分,按图 2.36 装妥。接通恒温水浴控制 25℃(恒温槽调节参见实验 2.1),样品管内装入蒸馏水,使水面刚与毛细管端面相切。注意毛细管务必与液体平面垂直。打开抽气滴液漏斗的活塞,让水缓慢滴下,使毛细管口逸出的气泡速度以 3~5s 一个为宜,记录压力计两侧最高和最低读数各三次,求取 Δh 的平均值,并同时记录温度。

同法测定各浓度正丁醇水溶液的 Δh 值。

五、数据处理

1. 由实验温度下水的表面张力算出毛细管常数 K。
2. 计算各溶液的表面张力 σ(表 2.22),并作 σ-c 曲线。
3. 在 σ-c 曲线上分别求浓度为 0.050、0.100、0.150、0.200 和 0.300mol·L^{-1} 的 (dσ/dc) 值,可按镜像法求取或用计算机处理(参见本书相关内容)。
4. 由上述各 (dσ/dc)$_x$ 值,用 Gibbs 公式计算各对应的 c/Γ 值,以 c/Γ-c 作图,从所得直线的斜率求取 Γ_∞。
5. 计算正丁醇分子的横截面积 S。

表 2.22 表面张力测定结果记录

样品	正丁醇浓度/mol·L^{-1}	Δh/cm	正丁醇表面张力
1	0.010		
2	0.020		
3	0.050		
4	0.100		
5	0.150		
6	0.200		
7	0.250		
8	0.300		
9	0.350		

六、注意事项

1. 要使毛细管端口刚好与液面相接触（为什么?）。
2. 安装仪器时应注意毛细管是否垂直。
3. 测定正丁醇溶液的表面张力时，浓度以由稀至浓为准。

七、思考题

1. 为何要使毛细管端口刚好与液面相接触？
2. 测定中如气泡逸出速度很快，甚至几个泡一起脱出，对测定结果有何影响？
3. 为什么要读取压差计上的最大压力差？

Experiment 2.12　Determination of Surface Tension of Solutions

Objectives

To measure, accurately, the surface tension of aqueous solutions of n-butyl alcohol and to calculate the cross section area of an adsorbed n-butyl alcohol molecule in a monolayer. Mastering the maximum bubble pressure method to measure the surface tension.

Theory/Principle

Solute molecules that lower surface free energy tend to be adsorbed at the surface. The extent of adsorption is given by the Gibbs equation,

$$\Gamma = -\frac{c}{RT}\frac{d\sigma}{dc} \tag{2.62}$$

Where Γ is excess surface adsorption amount, $mol \cdot m^{-2}$; R is gas constant, $8.314 J \cdot mol^{-1} \cdot K^{-1}$; T is absolute temperature, K; σ is surface tension, $N \cdot m^{-1}$; c is concentration in solution, $mol \cdot L^{-1}$.

A saturated adsorption monolayer was formed when the concentration in solution was increased to a certain amount. The adsorbed molecules occupied the whole surface.

The Fig. 2.37 is the σ-c curve of n-butyl alcohol solutions. Various $\frac{d\sigma}{dc}$ values can be obtained from the curve. Therefore, excess surface concentration, Γ can be calculated by the Gibbs Equation (2.62). The calculated Γ-c curve is Fig. 2.38. The saturated adsorption amount Γ_∞ can be determined from the curve. Supposed that the n-butyl alcohol formed a monolayer when it is under saturated adsorption condition, the cross section area of an adsorbed n-butyl alcohol molecule in a monolayer can be calculated by the following formula:

$$S = \frac{1}{\Gamma_\infty N_0} \tag{2.63}$$

S is the cross section area of n-butyl alcohol molecule and N_0 is the Avogadro constant. In fact it is difficult to obtain the saturated adsorption amount Γ_∞ from Γ-c curve. Here we used

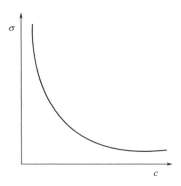

Fig. 2.37 the relationship of surface tension σ with concentration c

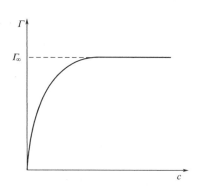

Fig. 2.38 excess surface adsorption Γ amount with concentration c

Langmuir constant temperature adsorption equation,

$$\frac{c}{\Gamma}=\frac{c}{\Gamma_\infty}+\frac{1}{K\Gamma_\infty} \tag{2.64}$$

to determine the saturated adsorption amount Γ_∞. In the above equation, c is the equilibrium concentration and K is a constant.

Through plotting $\frac{c}{\Gamma}$ versus c (from the experimental results), a line can be obtained and the slope of the line is the reciprocal of saturated adsorption amount Γ_∞.

The schematic diagram and principle of maximum bubble pressure method to measure surface tension is the followings (Fig. 2.39).

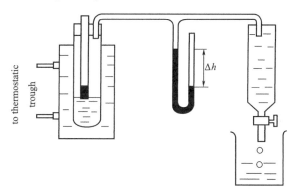

Fig. 2.39 The schematic diagram of the apparatus for measurement of surface tension by maximum bubble pressure method

The capillary end is set to cutting the surface of the liquid and it will rise along the capillary. Opening the piston of dropping funnel, and water will drip slowly to make the pressure of inner capillary larger than that of the liquid surface. When the acting force on the capillary end originated from this pressure difference is larger than the surface tension of capillary liquid slightly, the bubbles are extruded from capillary mouth. The maximum pressure difference p_{max} can be read from pressure gauge,

$$p_{max}=p_{air}-p_{sys}=\Delta h\rho g$$

Where Δh is the liquid column difference from U-type pressure gauge; ρ is the liquid density and g is the acceleration of gravity.

If r is the inner diameter of capillary, the acting force on the capillary end originated from this pressure difference is

$$F = \pi r^2 p_{max} = \pi r^2 \Delta h \rho g$$

The surface tension of the bubble at the mouth of the capillary is

$$F = 2\pi r \sigma$$

These two forces should be equal when the bubble is extruded,

$$\pi r^2 \Delta h \rho g = 2\pi r \sigma$$

$$\sigma = \frac{1}{2} r \Delta h \rho g \tag{2.65}$$

If the same capillary was used in the experiment, $\frac{1}{2} r \rho g$ is called the apparatus constant. We use K to represent it.

$$\sigma = K \Delta h \tag{2.66}$$

If we use a known surface tension sample as standard, the apparatus constant K can be obtained after Δh was measured from experiment. The surface tension σ of unknown sample can be determined using the same apparatus and method.

Apparatus and Chemicals Required

The surface tension measurement apparatus; nine volumetric flasks (100mL); n-butyl alcohol (AR).

Procedure

1. Preparation of various concentrations of n-butyl alcohol solutions

Preparing various concentrations (0.010, 0.020, 0.050, 0.100, 0.150, 0.200, 0.250, 0.300, 0.350 mol·L^{-1}) of n-butyl alcohol solutions according to general method in the laboratory.

2. Measurement of surface tension of the liquid

Assembling the apparatus of measuring surface tension according to the schematic diagram in the above text. Keeping the water temperature as 25 ℃ with constant-temperature water bath. Distilled water is enclosed in the sample tube. The capillary end is set to cutting the surface of the liquid. Attention: the capillary should be perpendicular to the liquid surface. Opening the piston of dropping funnel, then water will drip slowly. Keeping the velocity of bubbling at 1 bubbling per 3~5 seconds. Reading the highest and lowest value at the two sides of the U-type pressure gauge for three times and the average difference in height is Δh. Writing down the liquid temperature when reading difference in height.

Measuring the average difference in height of n-butyl alcohol using the same method.

Observations and Measurements

Table 2.23 Data record of the experiment

room temperature:　　　　　　surface tension of water:

apparatus constant K:

sample number	concentration of n-butyl alcohol/mol·L^{-1}	average difference in height Δh/cm	surface tension of n-butyl alcohol
1	0.010		
2	0.020		
3	0.050		
4	0.100		
5	0.150		
6	0.200		
7	0.250		
8	0.300		
9	0.350		

Calculations

1. Calculating the apparatus constant K from the water surface tension at the experimental temperature.

2. Calculating the surface tension of n-butyl alcohol solutions at various concentrations and plotting σ-c curve.

3. Calculating the $\left(\dfrac{d\sigma}{dc}\right)_r$ from σ-c curve at the concentrations of 0.050, 0.100, 0.150, 0.200, 0.300 mol·L^{-1}.

4. Calculating $\dfrac{c}{\Gamma}$ at the above concentrations using $\left(\dfrac{d\sigma}{dc}\right)_r$ and *Gibbs equation*. Plotting $\dfrac{c}{\Gamma}$ versus c and getting the saturated adsorption amount Γ_∞ from the slope of the line.

5. Calculating the cross section area of n-butyl alcohol molecule.

Notes

1. The capillary must be cleaned before use.

2. The capillary end must be set to cutting the surface of the liquid.

3. The sequence of measuring surface tension of n-butyl alcohol must be from dilute concentrations to dense ones.

Questions/Exercises

1. Why should keep capillary end be cut to the surface of the liquid?

2. How the measurement results will be if the bubbling velocity is too fast?

3. Why should read the largest difference of height in U-type pressure gauge?

实验 2.13　凝固点降低法测定溶质的摩尔质量

一、实验目的

掌握稀溶液凝固点下降法测定溶质的摩尔质量的原理及方法。

二、实验原理

含非挥发性溶质的二组分稀溶液（当溶剂与溶质不生成固溶体时）的凝固点将低于纯溶剂的凝固点。这是稀溶液的依数性之一，当指定了溶剂的种类和数量后凝固点降低值取决于所含溶质分子数目，即溶剂的凝固点降低与溶液的浓度成正比。以方程表示这一规律则有：

$$\Delta T_f = T_f^* - T_f = K_f b_B \tag{2.67}$$

式中，T_f^* 为溶剂的凝固点；T_f 为溶液的凝固点；K_f 为凝固点降低常数；b_B 为溶质 B 的质量摩尔浓度。因 $b_B = (g/G)M$，故式（2.67）可改写成：

$$M = K_f \frac{g}{G \cdot \Delta T_f} \tag{2.68}$$

式中，M 为溶质的摩尔质量，$kg \cdot mol^{-1}$；g 和 G 分别为溶质和溶剂的质量。若已知溶剂的 K_f 值，通过实验测得 ΔT_f 值，可利用式（2.68）计算溶质的摩尔质量。

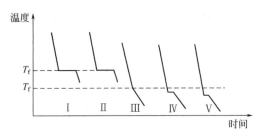

图 2.40 冷却曲线

各种溶剂具有不同的 K_f 值，苯的 K_f 为 $5.12\ K \cdot kg \cdot mol^{-1}$，其凝固点为 $5.51℃$。

纯溶剂的凝固点是它的液固相平衡的温度。若将纯溶剂逐步冷却，其温度随时间变化关系，即冷却曲线如图 2.40 Ⅰ 的形状，水平段对应的温度为凝固点。但实际过程中，液体在开始凝固前常出现过冷现象，即温度降至凝固点温度以下一定值后才开始出现固体，同时由于放热使温度回升至液固相平衡温度，待液体全部凝固后，温度再下降。实际纯溶剂冷却曲线如图 2.40 Ⅱ 的形状。

稀溶液的凝固点是液相混合物与溶剂的纯固相共存的平衡温度。若将溶液逐步冷却，其冷却曲线与纯溶剂不同，如图 2.40 Ⅲ、Ⅳ 所示。由于部分溶剂凝固而析出，使剩余溶液的浓度逐渐增大，剩余溶液与溶剂固相平衡温度也逐渐下降。因此，溶液冷却曲线不出现水平段，只出现转折点，冷却曲线 Ⅲ 的转折点对应的温度应为溶液的凝固点。实际溶液冷却过程也出现过冷现象，过冷后回升所达到的温度常较原始给定浓度的溶液凝固点低。这是由于溶液过冷后析出大量固体，使溶液浓度偏离原给定浓度而造成的。如冷却曲线 Ⅳ 所示，若溶液过冷程度不大，对测得凝固点影响较小。若过冷严重，测得凝固点明显偏低，如冷却曲线 Ⅴ。因此在测定过程中应设法控制适当的过冷程度，一般可通过控制冷浴温度、搅拌速度等方法来控制过冷程度。

因为稀溶液的凝固点降低值不大，所以温度的测量要用较精密的数字式贝克曼温度计。

三、仪器与试剂

凝固点测定仪一套，见图 2.41；数字式贝克曼温度计一台；0~50℃的1℃分刻度温度计1支；500mL 烧杯、称量瓶各1个；25mL 移液管1支；分析天平；苯、萘（AR）。

四、实验步骤

1. 准备工作

在玻璃缸（或保温瓶）中放入碎冰，再放入适量水，碎冰约占冰水总量的 1/2。在 500mL 烧杯中放入约 200mL 水和几块冰，其温度调至 5℃ 左右。用移液管取 50mL 苯于洗净干燥的内管中,将内管直接插入冰水浴中冷却,装好内管的搅拌器和塞子,2~3min后,液态苯中刚出现结晶时,插入仔细擦干的测温探头。

2. 凝固点测定

在插入测温探头后,内管中仍有苯的结晶,此时温度计读数为凝固点相对近似值。若结晶消失,可再插入冰水浴中,直到有结晶出现后,记录温度值。

3. 准确凝固点测定

取出内管,温热之(手握)。使苯的结晶全部熔化。再次将内管插入冰浴中,缓慢搅拌,使其全部冷却,并观察温度计读数,当苯液的温度降至高于近似凝固点 0.5℃ 时,取出内管,擦干水,移至外套管中,停止搅拌。待温度低

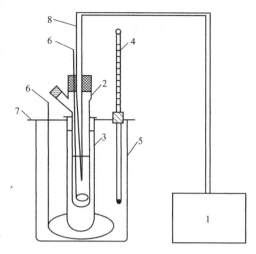

图 2.41 凝固点测定仪
1—数字式贝克曼温度计;2—内管;3—外套管;
4—普通温度计;5—玻璃缸;6—搅拌器;
7—盖板;8—测温探头

于近似凝固点 0.3℃ 左右时,急速搅拌(约 100 次·min^{-1})。大量结晶出现,温度开始回升。此时应改为缓慢搅拌。一直到温度达到最高点,此即纯溶剂的相对精确凝固点。重复测定三次,取平均值。

4. 溶液凝固点的测定

取出内管,温热之,使苯结晶熔化。准确称量 0.3g 左右萘。小心地将萘从支管加入内管中,按步骤 3 测准确凝固点 3 次。

5. 测定冷却曲线数据

在测最后两次溶液凝固点中,当温度冷至凝固点以上 1℃ 时,启动秒表,每隔 30s 读一次温度至温度达最低值,将回升前最低温度及相应的时间记下,温度回升过程中读 1~2 次温度及相应时间。当温度回升到最高点,记下最高温度及相应时间。过最高点后每隔 1min 记一次温度,连续读 4 次。

五、数据处理

1. 作溶液的冷却曲线,并解释冷却曲线每一步的物理意义。
2. 由式(2.68)计算萘的摩尔质量,苯的质量由其体积及室温下的密度求得。苯的密度与温度的关系为:$d = 0.90005 - 1.0636 \times 10^{-3} t - 3.76 \times 10^{-8} t^2 - 2.213 \times 10^{-9} t^3$。式中,$d$ 为密度,$g·cm^{-3}$;t 为温度,℃。
3. 计算相对误差。

六、思考题

1. 凝固点下降法测溶质摩尔质量使用范围如何?
2. 影响凝固点精确测量的因素有哪些?
3. 萘加入过多或过少对测定将带来什么影响?

Experiment 2.13　Determination of Solute Molar Mass by Freezing-point-depression Method

Objectives

1. To determine molar mass of naphthalene or osmotic pressure of the solution by freezing point depression method.

2. To master experimental technology of freezing point determination and understand deeply the colligative properties of solutions.

Principle

The colligative properties of a solution change in proportion to the concentration of solute dissolved in solvent. These properties depend only on the number of solute particles present in a given amount of solvent. At constant pressure, the pure liquid and pure solid can be in equilibrium at only one temperature, namely, the freezing point of the compound.

The change in the freezing point (ΔT_f) for a nonvolatile solvent can be determined using the following equation

$$\Delta T_f = \frac{R(T_f^*)^2}{\Delta H_m} \times \frac{n_2}{n_1 + n_2} \tag{2.69}$$

Where ΔT_f is the freezing point depression value; T_f^* is the freezing point of pure solvent; ΔH_m is the molar heat of solidification; n_1 and n_2 are amount of substance of solvent and solute respectively.

When the solution is very dilute, equation (2.68) becomes

$$\Delta T_f = \frac{R(T_f^*)^2}{\Delta H_m} \times \frac{n_2}{n_1} = \frac{R(T_f^*)^2}{\Delta H_m} \times M_1 m_2 = K_f m_2 \tag{2.70}$$

Where M_1 is the molar mass of the solvent, m_2 is the molality of the solute, and K_f is the freezing point depression constant of the solvent.

When the ΔT_f, K_f, weight of solvent W_1 and weight of solute W_2 are obtained, the molar mass of solute M_2 will be calculated using the following equation

$$M_2 = K_f \times \frac{W_2}{\Delta T_f \times W_1} \times 10^3 \tag{2.71}$$

The amount of freezing point depression value directly reflects the number of particles of solute in solution. When there is dissociation, association, solvation or formation of complex, etc., the apparent molar mass of the solute will be affected. Therefore, the freezing point depression of the solution can be used to study the ionization of electrolyte solution, the association degree of the solute and solvent permeability coefficient and activity coefficient.

Osmosis is the process whereby a solvent passes through a semipermeable membrane from low concentration to high concentration. Osmotic pressure is the pressure that must be applied on the high concentration side to stop osmosis. Important examples of semipermeable membranes are biological membrane, for example, the cell membrane and capillary wall of human body. So osmotic pressure is very important for injection, eye drops, and transfusion etc. Osmotic pressure is one of the colligative properties and can be expressed by osmolality.

Osmotic pressure is proportional to the molarity m of the solution, that is

$$\Pi = K_0 m \tag{2.72}$$

Where Π is osmotic pressure; K_0 is osmotic pressure constant. Since both the molarities m in Equation (2.70) and Equation (2.72) are the same, K_0 can be calculated as

$$K_0 = \frac{\Delta T}{K_f} \tag{2.73}$$

So osmotic pressure can be calculated by determining freezing point depression (ΔT_f) using the Equation (2.73).

The unit of osmolality is indicated by the mOsmole of solute in solution per kilogram of solvent, which can be calculated using the equation

$$\text{mosmole}(\text{mOsmol} \cdot \text{kg}^{-1}) = m \cdot i \cdot 1000$$

Where m is the molality, and i is the number of ions present per formula unit (e.g., $i=1$ for glucose; $i=2$ for NaCl or $MgSO_4$; $i=3$ for $CaCl_2$; $i=4$ for natrium citricum).

Osmole ratio is commonly used to evaluate osmotic pressure of preparation. One can determine osmole of test solution and 0.9% NaCl standard solution respectively (O_T and O_S) then calculate osmole ratio according to the formula

$$R = \frac{O_T}{O_S} \tag{2.74}$$

If a pure substance is heated to a liquid state and allowed to cool, the temperature of the liquid will begin to drop. When the freezing point of the substance is reached, the liquid begins to solidify and the temperature of the solid-liquid mixture will remain constant until all the liquid has been converted into solid. Then the temperature will begin to drop again. A graph of the temperature of such a system as function of time can be drawn and is called a cooling curve which exhibits a typical plateau at the freezing point of the substance (Fig. 2.42).

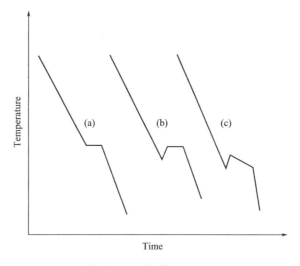

Fig. 2.42 Cooling curves

The cooling behavior of a solution is somewhat different from that of a pure liquid. The temperature at which the solution begins to solidify is lower (i. e. depressed) than that of the pure solvent. Additionally, there is a slow gradual fall in temperature as freezing proceeds [Fig. 2. 42 (c)]. The difference between the freezing points of the pure solvent and solution can be used to calculate the molality of the solution or the molar mass of the solute using Equation (2. 70) and Equation (2. 71).

In Fig. 2. 42 (b) and (c), the supercooling effect may be seen. Sometimes, a liquid or solution is solidified below the actual freezing point initially and then come back up to the freezing point temperature as the solid forms. This behaviour is commonly seen when very clean or new equipment is used because of lacking scratches or other irregularities that serve as sites for crystallization to begin. Supercooling may be overcome if adequate churning of the sample is provided. When determining the freezing point, the supercooling effect should be ignored if present.

Apparatus and Reagents

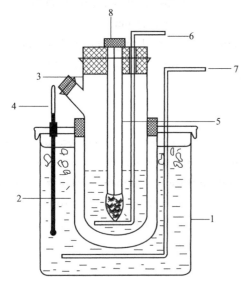

Fig. 2. 43 Experimental apparatus
1—large beaker; 2—air jacket; 3—test tube;
4—thermometer; 5—temperature probe;
6, 7—stirrer; 8—temperature measuring instrument

1. Apparatus: (for molar mass determination) freezing-point radiometer, thermometer, pipette (25 mL), tabletting machine, beaker, precision temperature measuring instrument; (for osmosis pressure determination) freezing-point radiometer, Beckmann thermometer, common thermometer, pipet, beaker.

2. Reagents: (for molar mass determination) cyclohexane (AR), naphthalene (AR); (for osmosis pressure determination) 1.0% sodium chloride, 0.9% sodium chloride injection, 5% glucose injection.

Procedure

1. Assemble the apparatus as shown in Fig. 2. 43 and adjust so that the stirrer, thermometer and probe will not touch each other and not touch the bottom and wall of the test tube. The test tube, probe and stirrer insider must be clean and dry.

2. Regulate the temperature 2~3 ℃ lower than the freezing point of cyclohexane and keep it basically unchanged through stirring and ice addition.

3. Place the test probe into the test tube and regulate the temperature measuring instrument to display a value as zero.

Procedures 4~6 designed for molar mass determination.

4. Determination of the reference freezing point of cyclohexane

Measure precisely 25 mL of cyclohexane with the pipette and pour it into a dry test tube. Stop the test tube with a rubber stopper to prevent evaporation of cyclohexane. Record the initial temperature of the cyclohexane. Place the test tube directly into ice-water bath with only the bottom of the tube in the bath. Keep stirring gently until the solid beings to form. Remove the test tube from the ice-water bath and dry the outside. Insert the test tube into the air jacket and stir the cyclohexane slowly but continuously. Record the temperature every 30 s until a constant temperature is obtained. Take this temperature as the reference freezing point of cyclohexane.

5. Determination of the freezing point of cyclohexane

Remove the test tube from the air jacket and warm it up till all solid has melted. Replace the test tube into the ice-water bath to cool the cyclohexane. Once the thermometer displays a temperature of approximately 0.5 ℃ higher than the reference freezing point of cyclohexane, remove the test tube from the bath and dry the outside. Place the test tube into the air jacket again and stir it slowly. When the temperature is about 0.2~0.3 ℃ lower than the reference freezing point, stir quickly to prevent a supercooling over 0.5 ℃. When the temperature begins to increase, stir slowly and record the temperature every 30 s until a constant temperature is obtained. This is the freezing point of cyclohexane. Repeat the cooling process for three times and the absolute average error must be less than 0.003 ℃. Save the test tube of cyclohexane for the next procedure.

6. Determination of the freezing point of the solution

Remove the test tube and allow it to warm until all the cyclohexane is melted. Add a sample tablet (about 0.2~0.3 g) into the test tube and stir the solution until the solid is completely dissolved. Determine the freezing point of the solution using the same procedure described in 4 and 5. Repeat the cooling process for three times and the absolute average error must be less than 0.003 ℃.

Procedures 7~8 designed for osmosis pressure determination.

7. Calibration of thermometer

Insert the sensing transmitter of Beckmann thermometer into the sample tube with 25 mL of 1.0% NaCl solution. Then put the sample tube into ice-water bath. Record the temperature with time variation at stirring until a constant temperature is obtained. Determine the freezing point of pure water using the same method. Thermometer can be calibrated through the two values.

8. Sample determination

(1) Insert the sensing transmitter of Beckmann thermometer into sample tube with 25 mL of 0.9% NaCl injection. Then put the sample tube into ice-water bath. Record the temperature with time variation at stirring until a constant temperature is obtained.

(2) The freezing point depression of 5% glucose injection is determined by the same method.

9. Dispose of the solution into the labeled waste container in the hood. Rinse the test tube, probe and stirrer with acetone and dispose these rinsings in the waste bottle.

Data Recording and Processing

1. Calculate the density of cyclohexane at room temperature using the equation

d (kg·m^{-3}) $= 0.7971 \times 10^3 - 0.8879t$, and then calculate the mass W (g) of cyclohexane to be taken.

2. Calculate the molar mass of naphthalene using Equation (2.71). ($T_f^* = 279.7$ K, $K_f = 20.1$ K·kg·mol^{-1})

3. Fill Table 2.24 and Table 2.25 with the data measured and calculated.

Table 2.24　Data record chart

solution	c	T_f/℃	ΔT_f/℃	thermometer calibration
standard 1				
standard 2				

Table 2.25　Data processing chart

solution	T_f/℃	ΔT_f/℃	mosmole/mOsmol·kg^{-1}
0.9% NaCl injection			
5% glucose injection			

Questions

1. What is the explanation of the formation of supercooling phenomena? How to control the degree of supercooling?

2. What would be the effect on the experiment result if too much or too less solute was added?

3. What would be the effect on the calculated molar mass if some impurity were present in naphthalene or some of the cyclohexane evaporated before the solute was added?

实验 2.14　二组分金属相图的测定

一、实验目的

用热分析法测绘锡-铋二元合金相图。

二、实验原理

金属的熔点-组成图可根据不同组成的合金的冷却曲线求得。将一种合金或金属熔融后，使之逐渐冷却，每隔一定时间记录一次温度，表示温度与时间的关系曲线称为冷却曲线或步冷曲线。当熔融系统在均匀冷却过程中无相的变化时，其温度将连续均匀下降，得到一条平滑的冷却曲线；如在冷却过程中发生了相变，则因放出相变热，使热损失有所抵偿，冷却曲线就会出现转折或水平线段，转折点所对应的温度，即为该组成合金的相变温度。对于简单的低共熔二元系统，具有图 2.44 所示的三种形状的冷却曲线，由这些冷却曲线即可绘出合金相图。如果用自动记录仪连续记录系统冷却的温度，则记录纸上所得的曲线就是冷却曲线（以 Bi-Cd 合金冷却曲线为例，如图 2.45 所示）。

用热分析法测绘相图时，被测系统必须时时处于或接近相平衡状态，因此系统的冷却速度必须足够慢才能得到较好的结果。

Sn-Bi 合金相图还不属于简单低共熔类型，当含 Sn 85% 以上即出现固熔体。因此用本实验的方法还不能作出完整的相图。

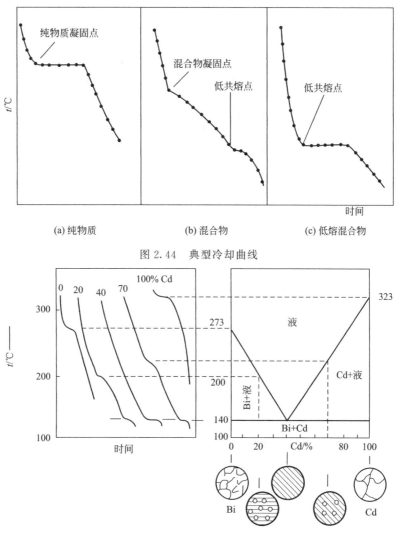

(a) 纯物质　　(b) 混合物　　(c) 低熔混合物

图 2.44　典型冷却曲线

图 2.45　Bi-Cd 合金冷却曲线及相图

三、仪器与试剂

镍铬-镍硅热电偶 1 支；XWC-100（或 XWC-200）型自动平衡记录仪 1 台（如无记录仪则用光点反射式检流计 1 台；钟表 1 只）；小保温瓶 1 个；盛合金的坩埚 5 个；电炉 2 个；调压器 2 个；坩埚钳 1 把；纯锡；纯铋；松香。

四、实验步骤

1. 配制样品

用感量为 0.1g 的台秤分别配制含铋量为 30%、58%、80% 的锡铋混合物各 100g，另外称纯铋 100g、纯锡 100g，分别放入 5 个 30mL 的瓷坩埚中。

2. 安装与调整自动记录仪

本实验所用的是 XWC-100 型或 200 型长图自动平衡记录仪（参见第 4 章 4.10）。图 2.46 中 1k 电位器作为信号电压 E_x 的分压器用。因本实验的记录仪全量程是 5mV。镍铬-镍硅热电偶在 300℃时的热电势约 12mV，所以要经分压器才能接到记录仪上。

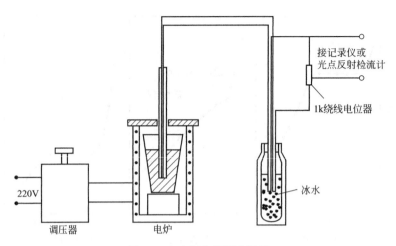

图 2.46　冷却曲线测定装置

使用 XWC-200 型记录仪时，为了使两组热电偶尽可能测得同一结果，除了制作热电偶的材料要一致外，在进行测量前，要将两组热电偶对同一温度进行测量，使它们在记录仪上指示同一数值。为此，可用水的沸点进行校正。将两热电偶热端浸入水中，加热到沸腾，调整记录纸速为 $300\text{mm} \cdot \text{h}^{-1}$。调整各自的分压器，使记录纸上的读数都等于 1.3mV。

3. 依次测纯铋、含铋 30%、58%、80%的锡铋混合物及纯锡的冷却曲线

方法如下：将装了样品的坩埚放入立式小电炉内，接通电炉电源，样品熔化后，在样品上面覆盖一层石墨粉或松香（防止金属被氧化），用小玻棒将熔融金属搅拌均匀。同时将热电偶热端插入熔融金属中心距坩埚底 1cm 处。样品温度不宜升得太高，一般在熔化全部金属后，再升高约 30℃ 即可停止加热，让样品在坩埚内缓缓冷却，同时开动记录仪，记录冷却曲线。冷却速度不能太快，最好保持降温速度在 $6\sim8\text{℃} \cdot \text{min}^{-1}$。当记录指示值小于 1.7mV 后，即可停止。

将已测之样品及另一待测样品同时加热熔化。从已测样品中取出热电偶插入待测的已熔化的样品中，同样方法用记录仪绘出冷却曲线。依次将全部样品测完。

如无记录仪，则用毫伏计或光点反射式检流计记录温度。每隔 30s 读一次数，过转折点后（注意合金有两个转折点）再读 4~5 次数，即停止。数据记录见表 2.26 和表 2.27。

表 2.26　实验数据记录

室温：　　　　　气压：

冷却过程中每隔 30s 检流计读数（用记录仪时不用此项）：

纯铋	
纯锡	
30%铋	
58%铋	
80%铋	

表 2.27　热电偶校正

组成	熔点/℃	记录仪读数
纯铋	271	
纯锡	232	
58%铋	139	
水	沸点	

五、数据处理

1. 以检流计读数为纵坐标，时间为横坐标，作出各合金的冷却曲线。
2. 用已知纯铋、纯锡、58%铋的熔点及水的沸点作标准温度。以冷却曲线上转折点的读数作横坐标，标准温度作纵坐标，做出热电偶的工作曲线。
3. 从工作曲线上查出 30%、80% 铋合金的溶点温度。以横坐标表示组成，纵坐标表示温度，作出 Sn-Bi 二元合金相图。

六、思考题

1. 金属熔融体冷却时冷却曲线上为什么会出现转折点？纯金属、低共熔金属及合金等的转折点各有几个？曲线形状为何不同？
2. 热电偶测量温度的原理是什么？为什么要保持冷端温度恒定？如何保持恒定？
3. 如果合金组成进入固溶体区（本相图含 Sn 85% 以上），则步冷曲线该是什么形状？

Experiment 2.14 Solid-liquid Binary Phase Diagram

Objectives

1. To construct Cd-Bi binary phase diagram from cooling curves.
2. To master basic principle and application of the thermocouple thermometer.

Principle

Binary eutectic phase diagram is widely used in steel smelting, petrochemical industry, ceramic technology, and product separation etc. A number of methods exist for constructing a phase diagram. One of the commonly used methods is by studying a cooling curve. Plot temperature versus cooling time, and the resulting cooling curve will show a plateau at a given temperature corresponding to the phase transition. By measuring a series of cooling curves for a range of compositions, the phase diagram of the tested system can be constructed from the temperature of phase transition (Fig. 2.47).

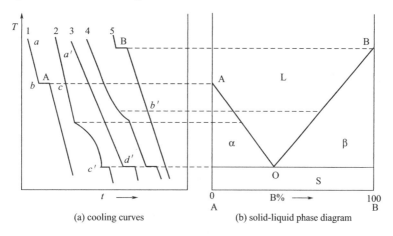

Fig. 2.47 Phase diagram drawn by cooling curves

If a pure substance is cooled down, the system cools relatively quickly until the melting point of the substance is reached, shown as curve 1 and curve 5. The slope of line ab depends on the extent of heat dispersion of the cooling system. The system exhibits two-phase equilibrium of solution and solid A at the melting point, and the temperature holds constant until all the liquid has been converted to solid. Then the temperature drops rapidly again.

Cooling curves of mixtures (e.g. curve 2, curve 4) are different from that of the pure substances. The temperature drops relatively quickly until point b' is reached. At point b', A is solidified and the system exhibits two-phase equilibrium. Because of the solidifying of A, the composition of liquid is continuously changed which resulting in reduced equilibrium temperature of A. However, the heat of fusion is released in this process and the temperature drops slowly (e.g. line $b'c'$). When the eutectic temperature is reached, component B starts to solidify, the system exhibits three-phase equilibrium and the temperature holds constant again until all the liquid has been converted to solid (e.g. line $c'd'$). Then the temperature drops rapidly again.

Curve 3 represents exactly the shape of mixture equal to eutectic composition, which is similar to the graphics of the pure material, but its horizontal section is a three-phase balance. Plotting temperature versus the composition of substance B (in mole fraction or weight percentage), the phase diagram can be constructed by connecting the transition temperature at specified composition.

Fig. 2.47 (b) is a simple eutectic phase diagram of two-component system. L represents the one phase region of liquid, α represents the two phase region of solid A+liquid, while β represents the two phase region of solid B+liquid, S represents the region of all solids of A+B. The closed line AOB is termed liquidus curve. At the horizontal line, three phases, that is liquid, crystals of A and B, all exist in equilibrium. Point 0 is the eutectic point. At this point also three equilibrated phases exist.

Apparatus and Reagents

1. Apparatus: 500W heating furnace, thermocouple, ceramic crucibles, potentiometer, rigid tubes, transformer.

2. Reagents: cadmium (AR), bismuth (AR), paraffin oil.

Procedure

1. 50g of each sample with different composition is prepared and is respectively put in the rigid tube. Concentrations of Cd in weight percentage are respectively 0%, 15%, 25% 40%, 55%, 75%, 90% and 100%. Then add about 3g of paraffin oil in each tube to prevent oxidation on air exposure of metal in heating.

2. Determine the cooling curves of samples in order of pure Cd, pure Bi, 90% Cd, 75% Cd, 55% Cd, 40% Cd, 25% Cd and 15% Cd. Place the hot end of the thermocouple with glass sheath into the rigid tube containing tested sample into the heating furnace (voltage controlled at about 160V), while the cold end of the thermocouple in ice-water to heat the sample above the liquid temperature. When all of the solids have been converted to liquids, stop heating. Record the temperature of the system every 1 min until the samples

have been completely solidified. Construct the cooling curve and determine the phase transition temperature. Use the melting points of pure Cd and pure Bi to calibrate the thermocouple. The melting points of Cd and Bi are 594.3 K and 544.6 K respectively.

Attention: control the cooling rates at 6~8 K·min^{-1}. The hot end of the thermocouple should be placed in the center of the sample and keep 1cm' distance from the bottom of the sample tube.

Data Recording and Processing

1. Construct the phase diagram of Cd-Bi binary system from cooling curves, and indicate the phase equilibrium in the designated regions.

2. Find the eutectic temperature and composition of the mixture.

Questions

1. What are the differences in horizontal section in cooling curves for samples with different compositions?

2. Whether can the heating curves be used to construct the phase diagram? Are there any other methods for construction of phase diagram?

第3章 提高型实验

实验 3.1 金属电极的制作、封装及电位测定

一、实验目的和要求

1. 学会金属电极的制作方法。
2. 学会电极的封装方法。
3. 了解万用表的使用方法,掌握电位的测试方法。

二、实验方法

1. 金属电极的制作

(1) 金属电极主要由用于制作电极的金属(如铁、铜、铝合金、镁合金等)、导线、焊锡等构成。

(2) 电极制作方法

① 取一底面积为 $1cm^2$ 的一圆柱形铁,用乙醇或丙酮洗去其表面的油,然后用蒸馏水冲洗,最后用吹风机将其吹干。

② 将铁电极的一横截面用砂纸打磨光亮,然后用焊锡将打磨光亮的一端与铜芯线焊接牢固。

③ 用万用表检测电极和导线间是否连通,如果连通则电极制作完成,否则重新进行焊接和检测,直到连通为止。电极如图 3.1 所示。

2. 电极的封装

图 3.1 电极示意图

(1) 剪一块合适大小的易拉罐,然后做成一圆柱形且一端封闭的圆柱桶,其内直径为 1.2~1.5cm。

(2) 将制作好的电极固定在圆柱筒中,保证电极在圆柱筒的中央。

(3) 用托盘天平称取 10g 环氧树脂(E44)、8g 对二甲苯、0.2g 乙二胺,然后将其混合均匀。

(4) 将混合均匀的环氧树脂溶液倒入离心管中,然后放入离心机中离心 3~5min,最后将离心好的环氧树脂溶液倒入圆柱筒中,放置 2h 固化后备用。

3. 溶液的配制

配制 1mol·L^{-1} 的 NaOH、NaCl 和 H$_2$SO$_4$ 的溶液。

4. 电位的测定

（1）洗净电解池，盐桥。

（2）将自制的研究电极端面打磨至镜面光亮度，然后依次用蒸馏水、丙酮和乙醇棉球擦拭。

（3）向电解池中加入 180mL 1mol·L^{-1}NaOH 溶液。

（4）分别在电解池中插入研究电极和参比电极，电解池如图 3.2 所示。

图 3.2 电解池示意图

（5）稳定 30min 后，用万用表测量其电位。

（6）每次实验完毕后，记下实验结果和研究电极外观及色泽，然后取出电极，用蒸馏水洗净电解池、电极和盐桥。

（7）重复步骤（2）～（6），分别测量研究电极在 1mol·L^{-1}NaCl 和 H$_2$SO$_4$ 溶液的电位。

三、注意事项

1. 测量电位前阅读万用表使用说明书，了解其使用方法。
2. 电极和导线间焊接一定要牢固。
3. 封装时一定要将焊接部位整体封装在环氧树脂中，且保证露出的端面和树脂间没有气泡和缝隙。
4. 电极表面一定要处理平整，光亮，干净。

四、实验结果和讨论

1. 实验结果

电解质溶液	电位	外观	色泽
NaOH			
NaCl			
H$_2$SO$_4$			

2. 对所制作的电极做全面的介绍和说明，并对封装提出自己的意见和建议。

五、思考题

1. 封装时，环氧树脂为什么要进行离心？
2. 实验所得的电位是平衡电位还是自腐蚀电位，二者有何不同，用草图说明。
3. 如果在电极端面上，金属和树脂之间有气泡或缝隙，对实验结果是否有影响？
4. 测量时，参比电极和研究电极间是断开的还是导通的？其原因为何？

实验 3.2 洗手液的研制及性能测定

一、实验目的和要求

了解洗手液的功能、配方设计和主要原料的作用及制备工艺，自行设计配方，拟定详细制备工艺条件和性能测定方法；掌握表面张力、pH 值、固含量等性能指标的测定原理和方法。

二、实验方法

1. 洗手液的研制

液体洗手液主要由水、表面活性剂、助洗剂、增稠剂、香精、色素等构成。

(1) 基本配方

原料	含量(质量分数)/%	主要作用
AES(70%)	8~12	去污
LAS(35)	10~15	去污
6501	2	助洗
甘油(丙三醇)	2	护肤
苯甲酸钠	0.5	防腐
NaCl	1~2	增黏
珠光剂	1	增色
香精	适量	
盐基玫瑰红	适量	调色
去离子水	至 100(73%~75%)	溶剂

注：AES——十二烷基聚氧乙烯醚硫酸钠；LAS——十二烷基苯磺酸钠；6501——椰子油烷基二乙醇酰胺。

(2) 配制方法

① 配制 100g 产品为基准。用 200mL 的烧杯，按配方要求用加量法分别称取 AES、LAS、苯甲酸钠。

② 在称好原料的烧杯中加入 60g 去离子水，在电炉上加热至 60~70℃，搅拌使原料全部溶解。

③ 适当降温后，用加量称样法加入 6501、甘油和珠光剂，搅拌均匀。

④ 加入余量去离子水、适当香精和盐基玫瑰红，搅拌均匀。

⑤ 搅拌下缓慢加入 NaCl，调节产品黏度。

2. 洗手液性能测定

(1) 水溶液表面张力的测定　用最大气泡法测定不同组成的洗手液水溶液的表面张力。配制产品的质量分数（%）组成如下：

$$1/5000、1/2000、1/1000、1/800、1/500$$

按照参考文献［1］所列装置和方法从低浓度到高浓度依次测定其 20℃时的表面张力。

(2) pH 值的测定　按照 GB 6383 以样品的 1%水溶液用精密 pH 试纸或 pH 计测定。用 pH 计测定所用仪器：温度计，精度 0.2℃；酸度计，精度±0.02pH；盘架天平，分度值 0.1g；烧杯，50mL。

（3）固含量测定　用重量法测定（详见参考文献 2）

$$固含量 = \frac{m_2}{m_1} \times 100\%$$

式中　m_1 为样品质量（2g，称准至 0.0002g）；m_2 为烘干后样品质量（105℃烘干 3h，称准至 0.0002g）。

（4）黏度测定　采用旋转黏度计（NDJ-1 型）测定。

（5）泡沫高度测定　若条件许可，按照 GB 7462 方法，用罗氏泡沫仪测定（参见参考文献 3）。亦可采用简化方法测定：取 1/1000 的溶液 80mL，加入 20mL 自来水，在 100mL 量筒中加入上述溶液 50mL，塞住量筒口上下摇动三次，记下泡沫高度。

三、实验结果讨论

1. 做溶液表面张力-组成图，估计产品的临界胶束浓度 CMC 值。
2. 给出产品的性能指标。参考格式如表 3.1 所示。

表 3.1　洗手液产品性能指标

指标名称	
外观	
色泽	
香型	
总活性含量	
pH 值	
固含量	
泡沫高度	
cmc	
黏度	

注：外观指聚集态、透明与否、是否黏稠等；总活性含量指总表面活性剂含量（包括 6501）；其余均为实测值。

3. 对所配制的产品作尽可能全面的介绍和说明，并提出改进产品配方的建议。

［教学要求］

按照教学指导书和参考文献，分组自拟性能测定实验计划，经老师审查后开始实验。

实验 3.3　氨基甲酸铵分解热力学函数的测定

一、实验目的和要求

用等压管法测定氨基甲酸铵的分解平衡压力，计算此分解反应热力学函数。

二、实验原理

氨基甲酸的分解平衡可用下式表示：

$$NH_2CO_2NH_4(s) \Longleftrightarrow 2NH_3(g) + CO_2(g)$$

此反应的标准平衡常数可表示为：

$$K^{\ominus} = \left(\frac{p_{NH_3}}{p^{\ominus}}\right)^2 \left(\frac{p_{CO_2}}{p^{\ominus}}\right) \tag{3.1}$$

式中，p_{NH_3} 和 p_{CO_2} 分别表示 NH_3 及 CO_3 的平衡分压；$p^{\ominus}$ 为热力学标准态的压力，等于 100kPa。设平衡总压为 p，因在等压管中 NH_3 及 CO_3 都是由氨基甲酸分解而来，$p_{NH_3} : p_{CO_2} = 2 : 1$。所以，在低压下，$p_{NH_3} = \frac{2}{3}p$，$p_{CO_2} = \frac{1}{3}p$。代入式（3.1）得：

$$K^{\ominus} = \frac{4}{27}\left(\frac{p}{p^{\ominus}}\right)^3 \tag{3.2}$$

因此，测得给定温度下的平衡总压 p，可以按式（3.2）算出分解反应标准平衡常数 $K^{\ominus}$。改变温度测得不同温度下的 $K^{\ominus}$。当温度变化范围不大时，$K^{\ominus}$ 与温度关系可用范特霍夫（Van't Hoff）等压方程式表示。下式为此方程的不定积分

$$\ln K^{\ominus} = -\frac{\Delta_r H_m^{\ominus}}{R}\frac{1}{T} + C \tag{3.3}$$

式中，$\Delta_r H_m^{\ominus}$ 为此反应的平均标准摩尔反应热；R 为通用气体常数；T 为热力学温度；C 为不定积分常数。根据不同温度 T 时 $K^{\ominus}$ 的实验值，可按式（3.3）求得实验温度范围内的平均标准摩尔反应热 $\Delta_r H_m^{\ominus}$。

在一定温度下，化学反应的标准摩尔吉布斯函变与标准摩尔熵变可用下列两个方程计算：

$$\Delta_r G_m^{\ominus} = -RT\ln K^{\ominus} \tag{3.4}$$

$$\Delta_r S_m^{\ominus} = \frac{\Delta_r H_m^{\ominus} - \Delta_r G_m^{\ominus}}{T} \tag{3.5}$$

三、仪器与试剂

仪器与装置如图 3.3 所示；氨基甲酸铵（实验室制备）；液体石蜡；蓖麻油。

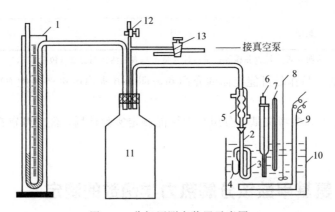

图 3.3　分解压测定装置示意图

1—U形压差计或精密压差计；2—等压管；3—盛液体石蜡部分；4—盛氨基甲酸部分；
5—冷凝器；6—接触温度计；7—精密温度计；8—搅拌器；9—电加热器；10—水浴；
11—缓冲瓶；12—二通活塞；13—三通活塞

四、实验步骤

1. 在洗净烘干的等压管 2 中装入少许氨基甲酸铵，使固体粉末集中到管 4 部位。在管 3 处装适量的液体石蜡（并加入 1~2 滴蓖麻油，作为消泡剂）。

2. 装好等压管，关闭活塞 12，启动真空泵。转动活塞 13，使真空泵接通装置，将压差计上汞柱高之差抽至 720mm 左右。转动活塞 13，切断装置，并使真空泵通大气，停止真空泵。检查系统是否漏气（压差计的汞柱高之差在 5min 内不发生变化，即认为不漏气）。

3. 启动恒温槽的搅拌器、电子继电器及电加热器，调节恒温槽的温度至 30℃。此步操作可与上一步试漏同时进行。

4. 若装置不漏气，可进行分解压测定。在试漏时由于系统维持 720mm 左右汞柱高之差的真空度，使氨基甲酸铵不断分解。分解得的 NH_3 及 CO_2 气体不断将管 4 部分空气带出。约 10min 后，打开活塞 12，缓缓放入空气，使等压管 3 处的液体石蜡对平，并达 2min 不变，读取 U 形压差计数值和温度计指示值。

5. 启动真空泵，转动活塞 13，连接装置。当汞柱高之差抽至 720mm 左右，旋转活塞 13，使系统封闭，停泵。再排等压管 4 部分的空气，约 5min。打开活塞 12，缓缓放入空气，使等压管中液体石蜡对平（仍需维持 2min）。读 U 形压差计、温度计指示值。在温度不变条件下，若压差计示柱高之差与步骤 4 的读数相差不超过 2mm，说明等压管 4 部分的空气基本干净，实验数据有效（以步骤 5 为标准）。

6. 调节恒温槽温度至 35.0℃，恒温数分钟后，从活塞 12 缓缓放入空气，使等压管中液面对平，且保持 2min 不变，读 U 形压差计及温度计指示值。然后用相同的方法，每隔 5℃ 测分解压一次，至 50.0℃，共得 5 组实验数据。注意：在升高恒温槽温度的过程中，可适当地由活塞 12 放入一些空气，使氨基甲酸铵不致分解过快。放入空气时要细心，不得使空气倒灌到等压管 4 部分中。万一发生倒灌，必须重排空气及检查空气是否排尽（方法参考步骤 4、5）。

7. 读大气压计指示值（见表 3.2）。

8. 实验完毕，将空气缓缓放入系统，至汞柱高差为零。停止搅拌，切断恒温槽电源。

表 3.2 分解压测定试验记录

室温：　　　　大气压：

编号	温度/℃	左支汞柱高 /mm	右支汞柱高 /mm	汞柱高之差 /mm	分解压/Pa	$K^{\ominus}$	$\ln K^{\ominus}$	$\frac{1}{T}/10^3 K^{-1}$

五、数据处理

1. 计算不同温度下氨基甲酸铵的分解压力及反应的标准平衡常数 $K^{\ominus}$。
2. 用作图法求平均标准反应热；并求得标准平衡常数与温度的关系式。
3. 计算 30℃ 时氨基甲酸铵分解反应的 $\Delta_r G_m^{\ominus}$ 和 $\Delta_r S_m^{\ominus}$。

六、思考题

1. 什么条件下才能用测总压的办法测定平衡常数？
2. 如何判定等压管中盛样上方的空气已抽尽？
3. 本实验数据处理结果中，$\Delta_r H_m^{\ominus}$、$\Delta_r G_m^{\ominus}$ 和 $\Delta_r S_m^{\ominus}$ 应有几位有效数字？

实验3.4 溶胶的制备及ζ电势的测定

一、实验目的和要求

1. 掌握化学凝聚法制备$Fe(OH)_3$溶胶的方法。
2. 观察溶胶的电泳现象并测定其ζ电势。

二、实验原理

溶胶是一种粒径为$10^{-9} \sim 10^{-7}$ m（$1 \sim 100$ nm）的固体粒子（称分散相），在液体介质（称分散介质）中形成的多相高分散物系。由于分散粒子的颗粒小，比表面积大，其表面能高，这就使得溶胶处于热力学不稳定状态，这是溶胶物系的特征，研究溶胶的形成、稳定与破坏，均需从该特征入手。如胶粒有相互聚结变成较大颗粒并发生聚沉的趋势。因此制备溶胶时，需要有稳定剂起稳定作用。

1. $Fe(OH)_3$溶胶的制备

制备溶胶的方法有多种，本实验采用化学凝聚法。该法是借助于一定的化学反应在适宜的反应条件下（如反应物浓度、溶剂、温度、pH值、搅拌等），使反应物呈过饱和状态，然后由分子分散状态逐步凝聚结合成胶体粒子而得到溶胶。

$Fe(OH)_3$溶胶是用$FeCl_3$溶液在沸水中进行水解反应制备而成。反应式如下：

$$FeCl_3 + 3H_2O \underset{}{\overset{沸腾}{\rightleftharpoons}} \underset{(红棕色溶液)}{Fe(OH)_3} + 3HCl$$

由于水解进行得不完全，溶液中还存在着少量Fe^{3+}和Cl^-，起着稳定剂的作用。由m个$Fe(OH)_3$分子聚集成的胶核选择性地吸附了Fe^{3+}，再由静电作用吸引了溶液中异电离子（Cl^-），形成紧密层，胶核加上紧密层称为胶粒，并吸引介质中的异电离子而形成扩散层，胶粒与扩散层一起构成胶团。$Fe(OH)_3$溶胶胶团的双电层结构示意图表示如下：

$$\underbrace{\{\underbrace{[Fe(OH)_3]_m \, n\,Fe^{3+}, \overbrace{(3n-x)Cl^-}^{紧密层}}_{胶粒}\}^{x+} \, \overbrace{\underset{B}{\overset{A}{|}}xCl^-}^{扩散层}}_{胶团}$$

AB是紧密层与扩散层的分界面（也称滑移面）。

事实上，在胶粒和胶团周围都吸附有溶剂分子，而形成一定的溶剂化层。

2. 电泳现象与ζ电势

从上述胶团的双电层结构可知，$Fe(OH)_3$溶胶的胶粒带正电荷。在无电场作用时，整个胶团是电中性的。但在外加电场作用下，胶粒则带着扩散层中少量溶剂分子向阴极移动，这种现象称为电泳。胶粒与周围介质作相对移动时的滑动面是位于紧密层与扩散层的分界处。实验证明，在滑动面处显示出一定的电势差，这个电势差只有当分散介质中的胶粒受外界电场作用而运动时才表现出来。故称为电动电势，通常也称为ζ电势。

ζ电势是表征胶粒特征的重要物理量之一，对研究胶体性质及解决胶体的稳定性问题都很有意义。

ζ电势可用下式计算：

$$\zeta = \frac{4\pi\eta}{DH}u \tag{3.6}$$

式中，u 为胶粒的移动速度，$u=\frac{s}{t}$，cm·s^{-1}；s 为在 t 秒钟内胶粒移动的距离，即胶粒与辅助液的界面移动的距离，cm；t 为实验观察的时间，s；H 为电势梯度，$H=\frac{V}{300l}$，绝对静电单位/cm；V 为直流电场的电压差，V；l 为两极间的距离，cm，是指 U 形管的导电距离；η 为分散介质的黏度，Pa·s；D 为分散介质的介电常数。

$$1\text{绝对静电单位}=300\text{V}$$

故式（3.6）可化为以下的具体形式：

$$\zeta = \frac{4\pi\eta ls}{DVt}300\text{V} \tag{3.7}$$

如果分散介质是水，则 $D=81$；不同温度时的 η 可查第 5 章 5.4 水的黏度。

式（3.7）是根据溶胶与辅助液的电导率相等而得到的。本实验近似地符合这条件，故可用以计算 ζ 电势。

三、仪器与试剂

U 形电泳管（图 3.4）1 支；蒸馏水洗瓶 1 个；石墨电极（或铂电极）2 支；250mL 烧瓶 1 个；镇流器（提供 110V 直流电源）1 台；10mL 量筒 1 个；稳压器 1 台；滴定管架 1 个；停表 1 只；电炉（1000W）及石棉网各 1 个；软尺获细铜线（量 U 形导电距离）1 条。10%FeCl$_3$ 溶液；蒸馏水；辅助液（H$_2$O∶HCl 体积比为 300∶1；或 0.001mol·L^{-1}KCl 溶液）。

四、实验步骤

1. 制备溶胶

取 100mL 蒸馏水倒入 250mL 烧杯中，加热至沸。用 10mL 量筒取 5mL 10% FeCl$_3$ 溶液，慢慢加入沸水中，并不断搅拌，待 FeCl$_3$ 溶液加完后，再煮沸 2min 便得红棕色 Fe(OH)$_3$ 溶胶，立即用冷水将溶胶冷至室温，留待观察溶胶的电泳现象和测定 ζ 电势。

2. 在 U 形电泳管 （图 3.4） 中用界面移动法观察电泳现象

电泳管下端连接一个活塞和带漏斗的狭长管，作为注入溶胶之用。

实验前将电泳管洗净，关闭活塞。从漏斗管中注入少许溶胶，慢慢打开活塞 A，用少许溶胶充填活塞内孔（注意尽量不让溶胶进入 U 形管内；若溶液进入 U 形管内，可用蒸馏水洗净），以便逐出活塞孔内的空气。然后关闭活塞，再将漏斗盛满溶胶，并在 U 形管内注入辅助液，至液面达刻度高约 1.5～2.0cm 处。辅助液为 H$_2$O∶HCl（300∶1，体积比）配制而成。将电泳管固定在管架上。于 U 形管两端插入石墨电极，使两电极端面等高。慢慢旋开活塞 A，使溶胶缓缓地上升进入 U 形管，但一定要保持溶胶与辅助液间的界面清晰，直至界面距电极端面约 2～2.5cm 处，关闭活塞 A。接着通电，并选择

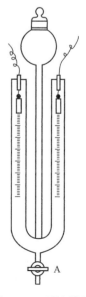

图 3.4 U 形电泳管

适当电压（约 50mV）。待界面明显清晰后，记录界面所对应的刻度数，同时开停表记录时间。15min 后停止通电。测出界面移动距离 s（取 U 形管两臂界面移动距离的平均值）、电极间距 l 和直流电压 V 后便可计算 ζ 电势。

五、注意事项

1. 电泳管活塞 A 孔内的空气务必逐尽。
2. 溶胶与辅助液界面应清晰。

六、数据处理

用式（3.7）计算 $Fe(OH)_3$ 溶胶的 ζ 电势。指出 $Fe(OH)_3$ 溶胶带何种电荷。

七、思考题

1. 何谓溶胶？化学凝聚法制备 $Fe(OH)_3$ 溶胶的基本原理是什么？
2. 何谓电泳？胶粒移动速度和哪些因素有关？
3. 何谓 ζ 电势？$Fe(OH)_3$ 溶胶的胶粒带何种电荷？
4. 电泳辅助液的选择根据什么条件？

实验 3.5　沉降分析

一、实验目的和要求

用沉降天平法测定碳酸钙粉末的粒度分布曲线。

二、实验原理

粒子在气体或液体介质中将受到重力作用而下沉。设粒子是球形的，则其受到的重力应为：

$$F_1 = \frac{4}{3}\pi r^3 (\rho - \rho_0) g \tag{3.8}$$

式中，r 为粒子半径；ρ、ρ_0 分别为粒子及介质的密度；g 是重力加速度。

粒子下沉时还同时受到摩擦阻力的作用。根据斯托克斯（Stokes）定律，粒子所受阻力为：

$$F_2 = 6\pi \eta r u \tag{3.9}$$

式中，η 为介质黏度；u 为粒子下沉速度。当重力和摩擦力达到相等时，粒子等速下沉，这时

$$6\pi \eta r u = \frac{4}{3}\pi r^3 (\rho - \rho_0) g$$

$$r = \frac{3}{\sqrt{2}} \sqrt{\frac{\eta u}{(\rho - \rho_0) g}} = k \sqrt{u} \tag{3.10}$$

由上式可见，当介质黏度、密度及粒子的密度为已知时，测得粒子的沉降速度以后，就可以计算出相应的粒子半径。

在 25℃时，$\rho_0 = 1.00 \times 10^3 \, kg \cdot m^{-3}$；$\rho_{CaCO_3} = 2.93 \times 10^3 \, kg \cdot m^{-3}$；$\eta = 8.90 \times 10^{-4} \, Pa$

·s，故 $k=4.88×10^{-4}$ m·$s^{1/2}$。若实验温度变化不大时，k 取此值。

分散体系由大小不同的粒子组成，为了得到分散体系的全部特征，常须测定大小不同的粒子的相对含量，即作出它们的分布曲线，这种分布曲线可由沉降曲线的图解处理求得。

沉降曲线以函数 $G=f(t)$ 表示，式中，G 是从实验开始经过时间 t 后所沉降的质量，或者是与此量成正比的其他物理量。

如果用扭力天平测出在时间 t 内粒子在介质中沉降到平盘上的粒子质量 G，以 G 对 t 作图即可得到沉降曲线。

设有五种大小不同的粒子，每种粒子单独沉降所得的沉降曲线如图 3.5 中曲线 1～5 所示。以曲线 3 为例，在到达时间 t_3 之前，粒子将均匀沉降。到 t_3 则所有粒子均沉降完毕，平盘质量保持 G_3 不变。t_3 是使所有在 h 高度内的粒子都完全沉降所需的时间，由此即可算出此粒子的沉降速度；

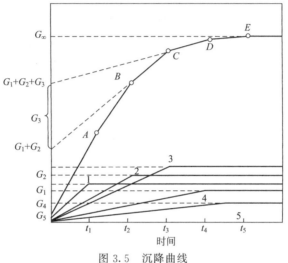

图 3.5　沉降曲线

$$\mu_3 = \frac{h}{t_3} \quad (3.11)$$

将 μ_3 代入式（3.10）即可求得此种粒子的半径 r_3。

当 $t<t_3$ 时，沉降曲线方程式是：

$$G = m_3 t$$

式中，m_3 是直线斜率。

当 $t>t_3$ 时，沉降曲线方程是：

$$G = G_3$$

如果样品中同时存在五种粒子，则变为图中上面一条沉降曲线。在任何时间曲线上的某一个点的沉降量，就相当于同时间五条线上相应点的沉降量之和。以线段 BC 为例，此线段上的任一点的沉降量是：

$$G = (m_3 + m_4 + m_5)t + G_1 + G_2 \quad (3.12)$$

线段 BC 与 t_2、t_3 间的沉降曲线相切，由式（3.12）的直线方程可知，其延长线与纵轴的交点即为 G_1+G_2，这就是在 t_2 时间已完全沉降的粒子量。线段 CD 的延长线与纵轴的交点代表 $G_1+G_2+G_3$，这两个交点之差就等于 G_3，即相当于半径为 r_3 的粒子量。

实际上粒子的分散度是很高的，其沉降曲线应是平滑的曲线。由上述分析很容易推广到这种情况。

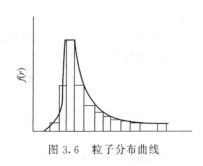

图 3.6 粒子分布曲线

为了作出粒子大小的分布曲线（图 3.6），需要求得分布函数 $f(r)$，用来表明半径 r 到 $r+\mathrm{d}r$ 之间的粒子质量占粒子总质量 G_∞ 的分数。

$$f(r)=\frac{1}{G_\infty}\times\frac{\mathrm{d}G}{\mathrm{d}r} \quad (3.13)$$

以 $\dfrac{\Delta G_i}{G_\infty \Delta r_i}$ 对半径 $r=\dfrac{r_i+r_{i+1}}{2}$ 作图，得梯状折线。根据折线形状可作出一条光滑的分布曲线，这曲线是 $f(r)$ 的近似图形，所取点愈多，近似程度愈高。

G_∞ 是沉降完毕后平盘上粒子的总质量。但由于细小粒子沉降很慢，需很长时间才能完成，故通常用作图外推法求 G_∞。

对沉降分析最大的干扰是液体的对流（包括机械的和热的原因引起的）和粒子的聚结。保持体系温度恒定可以减少热对流。添加适当的分散剂（多为表面活性物质）可防止粒子聚结。分散剂的类型和量必须经过试验，添加量一般不宜超过 0.1%，以免影响体系的性质。用于沉降分析的液体介质不应与粒子反应或使粒子溶解。其黏度和密度应与粒子密度结合起来考虑，使粒子有一定的沉降速度。

沉降分析只适合于大小在 $1\sim 50\mu m$ 范围内的颗粒。固体浓度不宜大于 1%，以保证粒子自由沉降。实际粒子往往并非球形，故测得的结果只能称为粒子的相当半径。

三、仪器与试剂

JN-A-500 型扭力天平（0～500mg，见图 3.7）1 台；玻璃沉降筒及恒温水夹套；停表 1 只；小天平；搅拌器；500mL、10mL 量筒各 1 个；400mL 烧杯 1 个；表面皿、牛角匙。碳酸钙试剂粉末；5% 焦磷酸钠溶液。

四、实验步骤

1. 为使 JN-A 型扭力天平适于作沉降分析用，需将原称量盒向上移动一定高度，盒底开槽，让平盘悬线通过。把底盘上的固定螺钉取掉一颗，使底座相对于天平转动一角度，才能使沉降筒正处于平盘吊钩下方，保持沉降筒有足够的高度。

调好天平的水平，打开开关 1，调整转盘 2，当天平达平衡时，平衡指针 4 应与零线重合，指针 3 的读数即为所称的质量。

2. 沉降玻筒直径约 55mm，高约 200mm，水夹套直径约为 100mm。平盘约 35mm，平盘边缘距筒壁不应小于 10mm。平盘用铝箔制成，用细镍铬丝作悬线，镍铬丝可在灯焰上烧红拉直。

沉降筒中装好煮沸冷却后的蒸馏水 300mL、5% $Na_4P_2O_7$ 溶液 6mL，将天平挂在天平吊钩 5 上，悬于沉降筒中，平盘距沉降底约 20mm。打开开关 1，转动 2，使指针 3 指零，打开 2 的调零盖，用螺丝刀转动调零螺钉，使平衡指针 4 与零线重合，同时从贴于沉降筒壁的标尺上

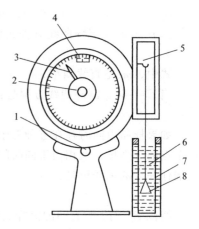

图 3.7　JN-A-500 型扭力天平
1—天平开关；2—指针转盘；3—指针；
4—平衡指针；5—平盘吊钩；6—沉降筒；
7—水夹套；8—平盘

读出平衡时平盘至水面的高度 h。然后取出平盘，记下水温。

3. 在台秤上称取约 1.5g 碳酸钙粉末，放在小表面皿上。将沉降筒中的水倒回量筒中，取少量水滴在表面皿上，用牛角匙仔细将聚集的粗粒碾散。然后用量筒中的水将其洗入烧杯中，再用量筒中的水使全部的粉末转入沉降筒里，用带橡皮薄圆盘的玻棒在量筒中上下抽动数分钟，使颗粒在整个液体中分布均匀后，迅速将沉降筒放在天平侧原位，将天平浸入筒内并挂在钩上，在小盘浸入 1/2 深度时打开停表，同时不断转动 2，使平衡指针 4 时时处于零线。一般可在 30s 内记下第一个读数。初期可每 30s 到 1min 记录一次天平读数，随沉降速度变慢，可适当延长读数间隔时间。直到间隔 5min 质量增加到 1mg 为止。

实验时应注意平盘处于沉降筒正中，盘底不能有气泡。

五、数据处理

1. 以沉降时间 t 为横坐标，沉降量 G 为纵坐标，作出光滑的沉降曲线。沉降量的极限值 G_∞ 可用作图法求得，即在沉降曲线纵轴左边作 $G\text{-}A/t$ 图（A 是任意常数，例如令 $A=1000$），由 t 值较大的各点作直线外推与纵轴相交，即为 G_∞。

2. 在沉降曲线上过适当的点用镜像法作切线交于纵轴，求得 ΔG_i，同时求得各点的沉降速度 u_i（$u_i = h/t_i$）和粒子半径 r_i。

将实验和计算结果列出表格，见表 3.3。

表 3.3 沉降分析实验数据

时间 t_i/s	沉降速度 $u_i/\text{cm}\cdot\text{s}^{-1}$	粒子半径 r_i/cm	r(平均) $=\dfrac{r_i+r_{i+1}}{2}$	Δr_i $=r_i-r_{i+1}$	ΔG_i	$f(r)$ $=\dfrac{\Delta G_i}{G_\infty \Delta r_i}$

3. 以 r（平均）对 $\dfrac{\Delta G_i}{G_\infty \Delta r_i}$ 作图，绘出粒度分布曲线。

六、思考题

1. 如果粒子不是球形的，则测得的粒子半径意义如何？如果粒子之间有聚结现象，对测得结果有何影响？

2. 粒子含量太多，或粒子半径太大或太小，对测定有何影响？

3. 什么原因会引起液体对流？什么原因会引起粒子聚结？如何减少它们对测定的影响？

实验 3.6　接触角的测定

一、实验目的

1. 掌握纯水及 0.1% 浓度的表面活性剂水溶液对表面的接触角的测定。
2. 熟悉 JJC-1 型接触角测定仪的结构与使用方法。

二、实验原理

接触角是一种固液界面的现象。通常规定在固体表面上的接触角（θ）为：在固-液-气三相接触线上任选一点 O，O 点作一垂直于三相接触线的平面。在此平面上，通过 O 点作一

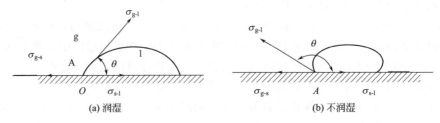

图 3.8 接触角示意图

气-液界面的切线,此切线与固-液交界线之间的夹角称为接触角(见图 3.8)。

接触角仪测定接触角时,先将被测液体由微量注射器取一小滴,滴在所需的固体平面上,再通过光学反光系统及放大系统将液滴放大于观测镜内,然后用镜头内的测角器测定其接触角值的大小。也可用照相机按一定时间间隔成像,在成像图片上测量不同时间的接触角,然后外推至 $t=0$ 处,所得接触角为前进接触角。

三、试剂仪器

JJC-1 型接触角测定仪,涂蜡玻璃片,表面活性剂溶液(0.1%)。

四、实验步骤

1. 调节水平仪,使仪器呈水平位置。
2. 接上电源,开启电源开关,调节亮度,使光线适度。
3. 调节前后移动旋钮、升降旋钮,使架板在显像板上有明显轮廓。
4. 把涂有石蜡的玻璃板放在架板上。
5. 用微量注射器将液滴滴在玻璃板上。利用调焦手轮和前后升降旋钮,使液滴在目镜分划板上清晰成像。
6. 调节刻度盘,使测角器的中心与液滴物像的任一边角顶点相重,而测角器的水平线则应与玻璃板的投影线相重合。
7. 调节镜头内测角器的标尺,测出液滴与平板相交处的切线,得出接触角的数值。注意作切线时,应尽可能取接近固体平板与液滴相交处焦点的切线,而不是一段弧的切线,否则所得接触角误差较大。
8. 用同样的方法测液滴物像另一端的接触角,重复三次并取平均值。

如用照相法测定,应洗出底片后再测量接触角。

五、数据处理

将测得的接触角列成表格,如用照相法测量,应对每一种样品在不同时间的接触角对时间作图,外推至 $t=0$ 处。

将所得结果与文献值相比较。

六、注意事项

对于含有表面活性剂溶液样品,测定时速度应尽可能快。

七、思考题

1. 在水中加入表面活性剂能使水与石蜡的接触角减小,原因何在?
2. 温度的变化及进样量的大小将各自如何影响接触角的测定值?

实验 3.7 B-Z 化学振荡反应活化能的测定

一、实验目的

1. 了解 Belousov-Zhahotinskii 反应（简称 B-Z 反应）的基本原理。
2. 掌握一般化学振荡反应的研究方法，初步认识体系在远离平衡态下的复杂行为。

二、实验原理

自然科学领域中普遍存在着非平衡、非线性的问题，这一新兴的研究领域受到了人们的重视，正在进行大量的研究工作。体系在远离平衡态下，由于本身的非线性动力学机制而产生宏观时空有序结构，Prigogine 等人称其为耗散结构（dissipative structure）。最经典的耗散结构是 B-Z 体系的时空有序结构，B-Z 体系是指由溴酸盐、有机物在酸性介质中，在有（或无）金属离子催化剂催化下构成的体系。它是由前苏联科学家 Belousov 发现，后经 Zhabot-inskii 发展而得名。

1972 年，R. J. Field，E. Koros，R. M. Noyes 等人认为：体系中存在着两个受 Br^- 浓度控制的过程 A 和 B；当 Br^- 浓度高于临界（crit.）浓度 $[Br^-]_{crit}$，时，发生 A 过程，当 Br^- 浓度低于 $[Br^-]_{crit}$ 时，发生 B 过程。也就是说：Br^- 浓度起着开关作用，它控制着 A 到 B 过程，再由 B 过程到 A 过程的转变。在 A 过程，由于化学反应 Br^- 浓度降低，当 Br^- 浓度达到 $[Br^-]_{crit}$ 时，B 过程发生。在 B 过程中，Br^- 再生，Br^- 浓度增加，当 Br^- 浓度达到 $[Br^-]_{crit}$ 时，A 过程发生。这样，体系就在 A 过程、B 过程间往复振荡。下面以 $BrO_3^-/Ce^{4+}/MA/H_2SO_4$ 体系为例说明。

当 Br^- 浓度足够高时，发生下列 A 过程：

$$BrO_3^- + Br^- + 2H^+ \xrightarrow{k_1} HBrO_2 + HOBr \tag{3.14}$$

$$HBrO_2 + Br^- + H^+ \xrightarrow{k_2} 2HOBr \tag{3.15}$$

其中反应式（3.14）是速率控制步骤，当达到准定态时，有 $[HBrO_2] = \dfrac{k_1}{k_2}[BrO_3^-][H^+]$，当 Br^- 浓度低时，发生下列 B 过程，Ce^{3+} 被氧化：

$$BrO_3^- + HBrO_2 + H^+ \xrightarrow{k_3} 2BrO_2 + H_2O \tag{3.16}$$

$$BrO_2 + Ce^{3+} + H^+ \xrightarrow{k_4} HBrO_2 + Ce^{4+} \tag{3.17}$$

$$2HBrO_2 \xrightarrow{k_5} BrO_3^- + HOBr + H^+ \tag{3.18}$$

反应式（3.16）是速率控制步骤，反应经式（3.16）、式（3.17）将自催化产生 $HBrO_2$。当达到准定态时，

$$[HBrO_2] = \dfrac{k_3}{2k_5}[BrO_3^-][H^+]$$

由反应式（3.14）和式（3.15）可以看出：Br^- 和 BrO_3^- 是竞争 $HBrO_2$ 的。当 $k_2[Br^-] > k_3[BrO_3^-]$ 时，自催化过程式（3.16）不可能发生。自催化是 B-Z 振荡反应中必不可少的步骤，否则，该振荡不能发生。Br^- 的临界浓度为：

$$[Br^-]_{crit} = \dfrac{k_3}{k_2}[BrO_3^-] \approx 5 \times 10^{-6}[BrO_3^-]$$

Br⁻的再生可通过下列过程实现：

$$4Ce^{4+} + BrCH(COOH)_2 + H_2O + HOBr \longrightarrow 2Br^- + 4Ce^{3+} + 3CO_2 + 6H^+ \qquad (3.19)$$

该体系的总反应为：

$$2H^+ + 2BrO_3^- + 2CH_2(COOH)_2 \longrightarrow 2BrCH(COOH)_2 + 2O_2 + 2H_2O \qquad (3.20)$$

振荡的控制物种是 Br⁻。

三、仪器与试剂

精密数字电压测量仪一套；丙二酸（AR）；硫酸铈铵（AR）；硫酸（AR）；溴酸钾（GR）；0.004 mol·L⁻¹ 的硫酸铈铵溶液；217 型甘汞电极（用 1 mol·L⁻¹ 的硫酸作液接）。

四、实验步骤

1. 用 1 mol·L⁻¹ 硫酸作 217 型甘汞电极液。
2. 按图 3.9 连接好仪器，打开超级恒温水浴，将温度调节至 25℃±0.1℃。

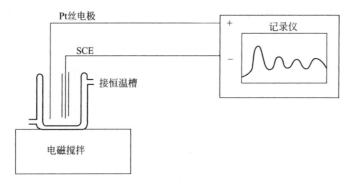

图 3.9 B-Z 反应实验装置

3. 配制 0.5 mol·L⁻¹ 丙二酸 250 mL、0.25 mol·L⁻¹ 溴化钾 250 mL、3.00 mol·L⁻¹ 硫酸 250 mL，在 0.2 mol·L⁻¹ 硫酸介质中配制 0.003 mol·L⁻¹ 的硫酸铈铵 250 mL。
4. 在反应器中加入已配好的丙二酸溶液、溴酸钾溶液、硫酸溶液各 15 mL。
5. 打开磁力搅拌器调节合适速度。
6. 将精密数字测量仪置于分辨率为 0.1 mV 挡（即电压测量仪的 2 V 挡），且为手动状态，甘汞电极接负极，铂电极接正极。
7. 恒温 5 min 后加入硫酸铈铵溶液 15 mL，观察溶液的颜色变化，同时开始计时并记录相应的变化电势。
8. 电势变化首次到最低时，记下时间 $t_{诱}$。
9. 用上述方法将温度设置为 30℃、35℃、40℃、45℃、50℃ 重复实验，并记下。

五、数据处理

测量诱导期（$t_{诱}$）和周期（T_1）随温度的变化。振荡的诱导期和周期的定义如图 3.10 所示。从加入硫酸铈铵到振荡开始定义为 $t_{诱}$，振荡开始后每个周期依次定义为 T_1，T_2，T_3，…。

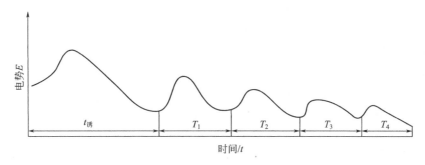

图 3.10　B-Z 反应的电势振荡曲线

六、思考题

1. 观察电势曲线与颜色和电势值的对应关系，分析 Pt 电极记录的电势。曲线主要反映哪个电对电势的变化？试说明理由。

2. 诱导期的长短与反应速率成反比，即：

$$1/t_{诱}=A\exp[-E_{表}/(RT)]$$

由此可得到：

$$\ln(1/t_{诱})=\ln A-E_{表}/(RT)$$

作 $\ln(1/t_{诱})$-$1/T$ 图，求出表观活化能 $E_{表}$。从 $\ln(1/t_{诱})$-$1/T$ 图得出的直线对诱导期中进行的反应有何推测？试说明理由。

3. 分析周期（T_1）随温度的变化（实验要求做几个温度下的振荡曲线，故有几个 T_1）。

实验 3.8　差热及热重分析

一、实验目的

1. 掌握差热和热重分析的基本原理，了解数据处理方法。
2. 对样品进行差热、热重分析，作出差热、热重分析谱图，并给予定性解释。
3. 了解差热-热重综合热分析仪的结构及测量原理，熟悉其使用方法。

二、实验原理

1. 差热分析（Differential Thermal Analysis，DTA）

差热分析是在程序控制温度下，测量试样与参比物（一种在测量温度范围内不发生任何热效应的物质）之间的温度差与温度关系的一种技术。

许多物质在加热或冷却过程中会发生熔化、凝固、晶型转变、分解、化合、吸附、脱附等物理化学变化。这些变化必将伴随体系焓的改变，因而产生热效应。其表现为该物质与外界环境之间有温度差。选择一种对热稳定的物质作为参比物，将其与样品一起置于可按设定速率升温的电炉中。分别记录参比物的温度以及样品与参比物间的温度差。以温差对温度（或时间）作图就可以得到一条差热分析曲线，或称差热谱图。

如果参比物和被测物质的热容大致相同，而被测物质又无热效应，两者的温度基本相同，此时测到的是一条平滑的直线，该直线称为基线。一旦被测物质发生变化，因此产生了热效应，在差热分析曲线上就会有峰出现。热效应越大，峰的面积也就越大。在差热分析中通常还规定，峰顶向上的峰为放热峰，它表示被测物质的焓变小于零，其温度将高于参比物。相反，峰顶向下的峰为吸收峰，则表示试样的温度低于参比物。差热分析中，温差信号是用两支相同的热电偶并联来测量的。两支热电偶分别置于样品及参比物中，由于样品及参比物之间有温差存在，就会在两支并联热电偶之间产生一个温差电势，一般这个温差电势很小。所以要经过放大电路放大以后，输入记录器或显示器，这就是差热分析仪测量原理。

热效应的大小是通过测定相对于某一参比物的温差来实现。这是由于所选用的参比物是一种在测定的变温范围内，其化学及物理性质非常稳定，不发生任何的物理变化及化学变化，当被测样品和参比物处于同步变温过程时，在某一温度，由于样品发生了物理或化学的变化，而引起了放热或吸热现象的发生，这时样品相对于参比样品有个温差出现，温差的大小取决于该步过程的热效应的大小，若将过程中的温差信号记录下来，就得到差热图谱如图3.11 所示。

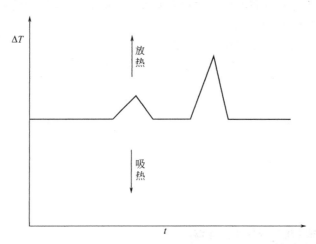

图 3.11　差热分析图谱示意图

差热曲线的峰形、出峰位置、峰面积等受被测物质的质量、热传导率、比热、粒度、填充的程度、周围气氛和升温速率等因素的影响。因此，要获得良好的再现性结果，对上述各点必须十分注意。一般而言，升温速度增大，达到峰值的温度向高温方向偏移；峰形变锐，但峰的分辨率降低，两个相邻的峰，其中一个将会把另一个遮盖起来。

2. 热重分析（Thermogravimetry，TG）

当被测物质在加热过程中有升华、汽化、分解出气体或失去结晶水时，被测的物质质量就会发生变化。这时热重曲线就不是直线而是有所下降。通过分析热重曲线，就可以知道被测物质在多少度时产生变化，并且根据失重量，可以计算失去了多少物质（如 $CuSO_4 \cdot 5H_2O$ 中的结晶水）。通过热重曲线我们就可以知道 $CuSO_4 \cdot 5H_2O$ 中的 5 个结晶水是分三步脱去的。图3.12 为热重装置结构示意图。

三、仪器与试剂

ZRY-1P 综合热分析仪，$CuSO_4 \cdot 5H_2O$（AR），$\alpha\text{-}Al_2O_3$（基准物）。

四、实验步骤

1. 打开热分析仪各单元电源（气氛、单元电源视使用情况而定），预热 20min 再进行操作。

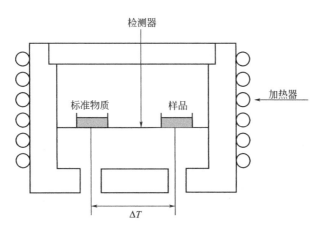

图 3.12 热重装置结构示意图

2. 打开电脑,双击桌面上的"ZRY-1P 应用软件"图标,然后点击对话框上的"采样"命令。

3. 设定各部分的量程参数,要求和热分析仪上的参数设置保持一致。

4. 装样

拧开石英玻璃管的上口,轻轻放下炉体。置有孔托盘于石英玻璃管上口。拧开白色瓷套,置于有孔托盘上。于白色瓷套上垫一纸片。在样品支架上,左边放参比物 Al_2O_3(一般不用换),右边根据被测样品的结束温度选择放置合适的坩埚(500℃以内选用 Al 坩埚,1000℃以内选用 Al_2O_3 坩埚)。取出垫片,套上白色瓷套。为防静电,可用湿布轻擦瓷套的外围。

装样前先调零:观察数据站接口单元的 TG 读数,用电减码调节此读数在 +00.000~+00.030 之间。电脑上点击"调零结束"命令(只可按一次)。

装样:退下瓷套,垫上垫片,取下坩埚装样(样品高度不要超过坩埚的 2/3)。通过增减样品量来调节数据站接口单元的 TG 读数在 03.200~04.500 之间。套上白色瓷套。

5. 在电脑的"采样"对话框中,设定样品的以下参数。

起始温度:一般在样品出峰前 50℃,但不要低于室温。

结束温度:必须小于 1000℃。

升温速率:10、15、20℃·min^{-1}(不要超过 20℃·min^{-1})。

保持时间:根据需要而定。(可以不填)

样品名:自定(必须输入)。

样品重:和数据站接口单元的 TG 读数保持一致。

气体:根据通入的气体(O_2 或 N_2)而定。(可以不填)

气体流量:和气氛单元的读数保持一致。(可以不填)

6. 拿掉炉体上的有孔托盘,升起炉体,先后旋紧固定螺帽和连接螺帽,同时打开电风扇。

7. 在电脑的"采样"对话框中,点击"确定"命令。同时双击打开桌面上"Balance"程序,设定温度程序,最基本的温度程序包括两个程序段,即升温程序和自然降温程序,设定参数如下。

(1) 升温程序段

初始温度:必须为 0。

终止温度:高于结束温度 100℃(必须小于 1100℃)。

速率:和升温速率保持一致。

时间:不用填。

点击"确定"命令。

（2）自然降温程序段

初始温度：和上面的终止温度保持一致。

终止温度：必须为0。

速率：不用填。

时间：－121（自然降温指令代码）。

点击"确认"命令→点击"完成"命令→温度程序有吗（是）→通信成功（确定）→点击"数据采集"命令→点击"采样"命令→点击"最小化"命令。

8. 启动电炉（按住电炉启动组上的绿色按钮），随后点击电脑桌面上的"RUN"命令。

9. 采样结束后，点击"存盘返回"命令，点击"STOP"命令。同时关闭电炉（按住电炉启动组上的红色按钮）。

10. 待温控单元上的温度读数小于200℃时，关闭仪器组的所有电源。

11. TG 数据处理

点击"热重数据处理"命令→点击"调用文件"命令→点击"处理设置"命令。根据TG曲线选择台阶个数，选定起始点位置（下降延迟前相对走平位置）和终点位置（下降延迟后相对走平位置）。点击"TG 处理"命令→点击"OK"命令（空）→是否重新测量（NO）。按相同步骤处理后面的台阶，处理完最后一个台阶后，按系统提示存盘。

12. DTA 数据处理

点击"差热数据处理"命令→点击"调用文件"命令→点击"处理设置"命令。根据DTA曲线选择峰个数，选定起始点位置（下降延迟前相对走平位置）和终点位置（下降延迟后相对走平位置）。点击"峰处理"命令→是否重峰（YES or NO）→点击"OK"命令（空）→是否重新测量（NO）。按相同步骤处理后面的峰，处理完最后一个峰后，按系统提示存盘。根据需要选择是否打印。

13. 数据转换

双击打开桌面上"数据转换"程序。选择仪器的类型（ZRY-1P）。点击"调用文件"命令，选择需要转换的文件。点击"数据转换"命令，按要求存盘。可以选择相应的软件（如Origin 软件）对数据进行处理。

五、注意事项

影响差热图谱的因素。差热图谱的特征受样品的处理，实验控制条件等因素的影响很大，稍不注意，则不能得到完整合格的图谱，主要影响因素有以下几个方面。

（1）样品的粒度要适中（一般在200目左右），并应和参比物的粒度保持一致，同时要求二者在样品管中的装填高度一致，以使二者的传热速度及热场均匀一致。参比物要求纯度高，并且使用前要在适当的温度下处理。置于干燥器内保存，严防吸水及其他污染。若样品的热效应大，可应用参比物进行稀释。

（2）测温热电偶应置于样品、参比物的中心位置，防止由此引起的基线漂移，影响峰的对称性，严防热电偶和样品管壁接触及外露。

（3）升温速度是非常重要的实验条件，若选择不当，直接影响图谱特征。升温速度过快，会使某些热效应小的峰不明显甚至丢掉；升温速度过慢，会使峰形变宽，在仪器噪声大的情况下，致使某些小峰不易辨认。不同的物质，需要选择不同的升温速度，不可一概而论。

(4) 本实验对于 $CuSO_4 \cdot 5H_2O$ 脱水过程选用 $10℃ \cdot min^{-1}$ 的升温速度较为合适。

六、数据处理

1. 记录实验条件。由峰的数目得到物质在被测温度内发生变化的次数。
2. 由各峰的位置知道每次变化发生的温度范围（即峰的起始温度到峰的结束温度）。用外推法从各差热曲线上确定起始反应温度。
3. 由出峰的方向的正负性确定该变化过程是吸热还是放热，一般规定放热峰为正，吸热峰为负。
4. 由各峰的峰面积大小，估算每个变化过程的热效应的大小。
5. 选择相应的软件（如 Origin 软件）对数据进行处理。

七、思考题

1. 影响差热分析实验结果的因素有哪些？如何防止？
2. 为什么要控制升温速率？升温过快有何后果？
3. 为什么差热分析中的温度必须从参比物中得到？
4. 在热重分析过程中，样品的重量为什么会发生变化？样品的重量有可能增加吗？

实验 3.9　金属腐蚀行为的电化学研究

一、实验目的

1. 测量铁在不同溶液中的极化曲线。
2. 了解金属的钝化现象及其他因素对它的影响。
3. 确定腐蚀电位、腐蚀电流、钝化电位、缓蚀效率等电化学参数。
4. 掌握 DJS-292 恒电位仪的使用方法。

二、实验原理

1. 金属的腐蚀

铁在 H_2SO_4 溶液中，将不断被溶解，同时产生 H_2，即：

$$Fe + 2H^+ \rightleftharpoons Fe^{2+} + H_2\uparrow \tag{a}$$

Fe/H_2SO_4 体系是一个二重电极，即在 Fe/H^+ 界面上同时进行两个电极反应：

$$Fe \rightleftharpoons Fe^{2+} + 2e \tag{b}$$

$$2H^+ + 2e \rightleftharpoons H_2 \tag{c}$$

反应(b)、(c) 称为共轭反应。正是由于反应（c）存在，反应（b）才能不断进行，这就是铁在酸性介质中腐蚀的主要原因。

当电极不与外电路接通时，其净电流 $I_{总}$ 为零。在稳定状态下，铁溶解的阳极电流 $I_{(Fe)}$ 和 H^+ 还原出 H_2 的阴极电流 $I_{(H)}$，它们在数值上相等但符号相反，即：

$$I_{总} = I_{(Fe)} + I_{(H)} = 0 \tag{3.21}$$

I_{Fe} 的大小反映 Fe 在 H^+ 中的溶解速率，而维持 $I_{(Fe)}$，$I_{(H)}$ 相等时的电势称为 Fe/H^+ 体系的自腐蚀电势 E_{corr}。

图 3.13 是 Fe 在 H^+ 中的阳极极化和阴极极化曲线图。当对电极进行阳极极化（即加更

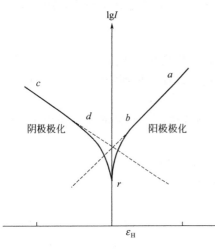

图 3.13 Fe 的极化曲线

大正电势）时，反应（c）被抑制，反应（b）加快。此时，电化学过程以 Fe 的溶解为主要倾向。通过测定对应的极化电势和极化电流，就可得到 Fe/H^+ 体系的阳极极化曲线 rba。由于反应（c）是由扩散步骤所控制，所以符合塔菲尔（Tafel）半对数关系，即：

$$\eta_{Fe} = a_{Fe} + b_{Fe}\lg[I_{(Fe)}/A\cdot cm^{-2}] \quad (3.22)$$

直线的斜率为 b_{Fe}。

当对电极进行阴极极化，即加更负的电势时，反应（b）被抑制，电化学过程以反应（c）为主要倾向。同理，可获得阴极极化曲线 rdc。由于 H^+ 在 Fe 电极上还原出 H_2 的过程也是由扩散步骤所控制，故阴极极化曲线也符合塔菲尔关系，即：

$$\eta_H = a_H + b_H\lg[I_{(H)}/A\cdot cm^{-2}] \quad (3.23)$$

当把阳极极化曲线 abr 的直线部分 ab 和阴极极化曲线 cdr 的直线部分 cd 外延，理论上应交于一点（z），z 点的纵坐标就是 $\lg[I_{corr}/A\cdot cm^{-2}]$，腐蚀电流 I_{corr} 的对数，而 z 点的横坐标则表示自腐蚀电势 E_{corr} 的大小。

2. 金属的钝化

当阳极极化进一步加强时，铁的阳极溶解进一步加快，极化电流迅速增大。当极化电势超过 ε_p 时，$I_{(Fe)}$ 很快下降到 d 点，如图 3.14 所示。此后虽然不断增加极化电势，但 $I_{(Fe)}$ 一直维持在一个很小的数值，如图中 de 段所示。直到极化电势超过 1.5V 时，$I_{(Fe)}$ 才重新开始增加，如 ef 段示。此时 Fe 电极上开始析出氧气。从 a 点到 b 点的范围称为活化区，c 是临界钝化点，从 c 点到 d 点的范围称为钝化过渡区，从 d 点到 e 点的范围称为稳定钝化区，从 e 点到 f 点称为超钝化区。ε_p 称为钝化电势，I_p 称为钝化电流。

铁的钝化现象可作如下解释：图 3.14 中 ab 段是 Fe 的正常溶解曲线，此时铁处在活化状态。bc 段出现极限电流是由于 Fe 的大量快速溶解。当进一步极化时，Fe^{2+} 与溶液中 SO_4^{2-} 形成 $FeSO_4$ 沉淀层，阻滞了阳极反应。由于 H^+ 不易达到 $FeSO_4$ 层内部，使 Fe 表面

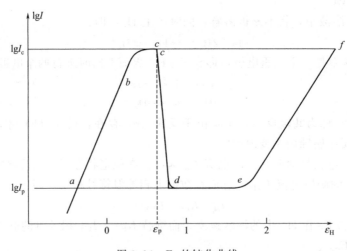

图 3.14 Fe 的钝化曲线

的 pH 值增加；在电势超过 0.6V 时，Fe_2O_3 开始在 Fe 的表面生成，形成了致密的氧化膜，极大地阻滞了 Fe 的溶解，因而出现了钝化现象。由于 Fe_2O_3 在高电势范围内能够稳定存在，故铁能保持在钝化状态，直到电势超过 O_2/H_2O 体系的平衡电势（+1.23V）相当多时（+1.6V），才开始产生氧气，电流重新增加。

金属钝化现象，在实际中有很多应用。金属处于钝化状态，这对于防止金属的腐蚀和在电解中保护不溶性的阳极是极为重要的。而在另一些情况下，钝化现象却十分有害。如在化学电源、电镀中的可溶性阳极等，则应尽力防止阳极钝化现象的发生。

凡能促使金属保护层破坏的因素都能使钝化后的金属重新活化，或能防止金属钝化。例如，加热、通入还原性气体、阴极极化、加入某些活性离子（如 Cl^-）、改变 pH 值等均能使钝化后的金属重新活化或能防止金属钝化。

对 Fe/H_2SO_4 体系进行阴极极化或阳极极化（在不出现钝化现象情况下）既可采用恒电流方法，也可以采用恒电势的方法，所得到的结果一致。但对测定钝化曲线，必须采用恒电势方法，如采用恒电流方法，则只能得到图 3.14 中 *abcd* 部分，而无法获得完整的钝化曲线。

恒电势方法和恒电流方法的简单线路如图 3.15 所示。

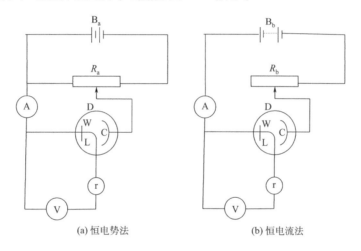

图 3.15　恒电势和恒电流测量原理

B_a—低压稳压电源；B_b—稳压电源；R_a—低电阻；R_b—高电阻；A—精密电流表；
V—高阻抗毫伏计；L—鲁金毛细管；W—工作电极；C—辅助电极；r—参比电极

3. 腐蚀速度及缓蚀效率

金属的腐蚀是电化学过程，在介质中发生金属的阳极溶解和介质中某组分的阴极还原：

阳极反应：$$M \longrightarrow ne + M^{n+}$$

阴极反应：$$O + ne \longrightarrow R$$

总反应：$$M + O \longrightarrow M^{n+} + R$$

式中，O 为氧化剂；R 为还原剂。

在多数场合，阴极反应通常为

$$\frac{1}{2}O_2 + H_2O + 2e \longrightarrow 2OH^-$$

或

$$H^+ + e \longrightarrow \frac{1}{2}H_2$$

金属作为阳极而被腐蚀时，失去的电子越多，则金属溶解得越多，失去的电子由阴极反应吸收。当阴极反应和阳极反应这一对共轭反应达到稳定状态时，可以得到一个稳定电位，即腐蚀电位，此时所对应的阳极电流密度或阴极反应电流密度即为腐蚀电流密度。

当加入缓蚀剂后，上述腐蚀反应中的阴极反应或阳极反应或两者的速度会减缓，其减缓的程度即为缓蚀效率：

$$缓蚀效率 = \frac{加缓蚀剂前腐蚀速度 - 加缓蚀剂后腐蚀速度}{加缓蚀剂前腐蚀速度} \times 100\%$$

腐蚀电流密度的测量可以采用动电位扫描法获得。即给电极加上一个线性变化的电位，同时记录电流密度随电极电位的变化。在强极化区，电极电位与电流密度的对数服从 Tafel 关系：$\eta = a + b\lg i$，将直线段外推至与 $\eta = 0$ 的直线相交，其交点处的值即为腐蚀电流的对数值。如图 3.16 所示。

三、仪器与试剂

研究电极（铁电极）、辅助电极（铂电极）、参比电极（饱和甘汞电极）；水磨砂纸、金相砂纸；电吹风。盐酸、硫酸、丙酮（棉球）、乌洛托品。万能胶，环氧树脂。DJS-292 恒电位仪；电解池如图 3.17 所示。

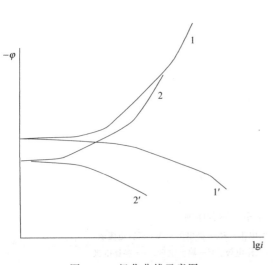

图 3.16 极化曲线示意图
1，1′— 未加缓蚀剂；2，2′—加缓蚀剂
1，2—阴极极化曲线，1′，2′—阳极极化曲线

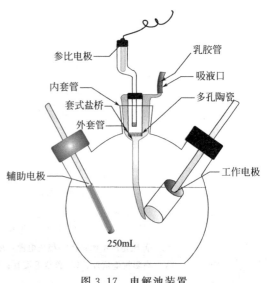

图 3.17 电解池装置

四、实验步骤

1. 制作研究电极

（1）将铁圆柱一端与铜芯线焊牢，柱面用砂纸打磨光，然后用丙酮去油、蒸馏水洗，吹干。

（2）将环氧树脂与固化剂按 10∶1 比例混合均匀。

（3）将上述铁块柱面及与铜芯线相连的一端用调制好的环氧树脂封好，放置 24h 固化后备用。

2. 配制溶液

(1) 1mol·L^{-1} H$_2$SO$_4$ 及 0.1mol·L^{-1} H$_2$SO$_4$；

(2) 1mol·L^{-1} HCl；

(3) 10%的乌洛托品。

3. 电化学测量

(1) 洗净电解池、盐桥、辅助电极，研究电极端面打磨至镜面光，丙酮棉球擦拭。

(2) 测铁的腐蚀速度及缓蚀剂缓蚀效率

① 在电解池中加入 200mL 1mol·L^{-1} HCl 水溶液。

② 插入三个电极，连接好线路。

③ 用恒电位仪测量开路电位（即腐蚀电位）。

④ 选择 LSV（线性扫描伏安法）极化方法，电位扫描速度设置为 0.5mV·s^{-1}，扫描范围为腐蚀电位正负 300mV。开始极化。

⑤ 实验完毕后，保存好结果。取出电极，将电解池、电极及盐桥洗净。

⑥ 在电解池中加入 200mL 1mol·L^{-1} HCl 水溶液及 10%乌洛托品 20mL，重新处理研究电极，重复步骤②～⑤。

(3) 测量铁的钝化曲线

①～③步与（2）同（溶液分别换为 1 mol·L^{-1} H$_2$SO$_4$ 及 0.1mol·L^{-1} H$_2$SO$_4$）

④ 选择 LSV 极化方法，电位扫描速度设置为 1mV·s^{-1}，扫描范围从腐蚀电位开始至出现超钝化。

⑤ 实验完毕后，保存好结果。取出电极，将电解池、电极及盐桥洗净。

五、数据处理

1. 分别求出 Fe 电极在不同浓度的 H$_2$SO$_4$ 溶液中的自腐蚀电流密度、自腐蚀电位、钝化电流密度及钝化电位范围。

2. 分别计算 Fe 在 HCl 及含缓蚀剂的 HCl 介质中的自腐蚀电流密度及按下式换算成腐蚀速率：

$$v = \frac{3600Mj}{nF}$$

式中，v 为腐蚀速率，g·m^{-2}·h^{-1}；j 为自腐蚀电流密度，mA·cm^{-2}；M 为 Fe 的摩尔质量，g·mol^{-1}；F 为法拉第常数，C·mol^{-1}；n 为发生 1mol 电极反应得失电子的物质的量。

3. 求出乌洛托品对铁在 1mol·L^{-1} HCl 溶液中的缓蚀效率。

六、注意事项

1. 测定前仔细阅读仪器使用说明书，了然仪器的使用方法。

2. 电极表面一定要处理平整、光亮、干净，不能有点蚀孔，这是实验成败的关键。

七、思考题

1. 平衡电极电位、自腐蚀电位有何不同？

2. 分析 H$_2$SO$_4$ 浓度对 Fe 钝化的影响。比较盐酸溶液中加和不加乌洛托品 Fe 电极上自腐蚀电流的大小。Fe 在盐酸溶液中能否钝化，为什么？

实验 3.10 粉末润湿性能的测定

一、实验目的

测定二氧化硅粉末-液体体系接触角，评估其润湿性能。了解 PC-01 粉末接触角测定仪的结构与使用方法。

二、实验原理

润湿是固体表面上的一种流体被另一种流体所取代的过程，它涉及三个相，而且其中两相为流体。通常所说的润湿是指固体表面上的气体被液体所取代的过程。

在理论上，可将润湿分成沾润、浸润和铺展三类。在恒温恒压下，这些过程的体系吉布斯函数的变化 $\Delta G_{T,p} = W'_r$，三类过程的功如下：

沾润 $$W_a = \sigma_L(\cos\theta + 1) \tag{3.24}$$

浸润 $$W_i = \sigma_L \cos\theta \tag{3.25}$$

铺展 $$W_s = \sigma_L(\cos\theta - 1) \tag{3.26}$$

式中，W_a、W_i、W_s 分别为沾润功、浸润功和铺展功；σ_L 为液体的表面张力；θ 为接触角。根据润湿过程发生的热力学条件 $W = \Delta G_{T,p} < 0$，由上述诸式得知，W 值与实现润湿的种类、σ_L 及 $\cos\theta$ 有关。σ_L 为正值，在指定条件下，W 值的正负号仅取决于 $\cos\theta$。因此，常用接触角（θ）来表征固体表面的润湿性能。

粉末接触角的测定可分为静态法和动态法。后者以 Washburn 方程为基础，逐步由高度平方法（h^2-t）发展到浸湿速度法（微分法）和重量平方法（m^2-t）。

根据 Washburn 方程，粉末柱体被液体浸湿的高度平方（h^2）与时间（t）有如下的关系式：

$$h^2 = \frac{C\bar{r}\sigma_L\cos\theta}{2\eta}t = Kt \tag{3.27a}$$

$$K = \frac{C\bar{r}\sigma_L\cos\theta}{2\eta} \tag{3.27b}$$

式中，σ_L 为液体的表面张力；η 为液体的黏度；θ 为液体对粉末的接触角；$\bar{r}$ 为粉末柱体之毛细管平均半径；C 为毛细管形状系数；$(C\bar{r})$ 为表观半径；K 为 h^2-t 直线斜率。

在实际测量中，确定表观半径（$C\bar{r}$）的常用方法是依据实验结果，选择某液体对此粉末完全润湿即 $\theta = 0$，$\cos\theta = 1$，h^2-t 直线斜率求算表观半径，对指定体系而言，$(C\bar{r})$ 值是恒定的。

确定了表观半径（$C\bar{r}$）值，可由 h^2-t 直线斜率计算其他液体对粉末的接触角（θ）。

三、仪器与试剂

PC-01 粉末接触角测定仪；刻度玻璃管（内径约 0.46cm，长 18cm）；甲苯（CP）；乙醇（CP）；丁醇（CP）；SiO_2 粉末。

四、实验步骤

1. 将样品管取出洗净干燥，精确测量样品管的内径。然后用待测液体湿润样品管壁，再干燥，用金属卡圈将微孔隔膜（可用滤纸）封闭样品管的底端。

2. 精确称取一定量的二氧化硅粉末，装入样品管内，振动夯实到某刻度，计算粉末柱体的堆积密度。

3. 在储液皿中注入待测液体，并小心放入仪器中央。然后将样品管连同仪器顶盖扣在仪器上端，上紧。调整样品管的位置和读数放大镜的位置，校核水准泡，插入温度计。

4. 转动微调螺旋，使样品管下端没入待测液体内，并同时启动秒表计时。液体穿过微孔隔膜向粉末柱体浸透，如图 3.18 所示。观察并记录液体浸湿粉末柱体的高度位置及时间。

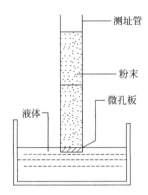

图 3.18 液体浸湿粉末柱体示意图

五、数据处理

1. 绘制二氧化硅-液体的 h^2 与 t 的直线关系图。
2. 计算表观半径 $C\bar{r}$。
3. 计算各液体对二氧化硅的接触角。

六、注意事项

样品管中所装样品的均匀性非常重要，故每次装样所用条件应尽可能保持一致。

七、思考题

1. 影响接触角的因素有哪些？
2. 如何用接触角评价粉末的润湿性能？

实验 3.11　活性炭比表面积的测定

一、实验目的

1. 测定活性炭在醋酸水溶液中对醋酸的吸附，推算活性炭的比表面积。
2. 使用表面积分析仪测定活性炭对氮气的吸附，计算活性炭的比表面积。
3. 比较两种测定方法的原理及结果，并分析原因。

二、实验原理

活性炭是用途广泛的吸附剂，可用于吸附气体，也可用于对溶液中某种物质的吸附。活性炭在水溶液中对不同的吸附质有不同的吸附能力，根据这种吸附作用的选择，在工业上有着广泛的应用，如糖的脱色提纯等。

吸附能力的大小常用吸附量 Γ 表示，Γ 通常指单位质量的吸附剂上吸附溶质的量。在恒温下，吸附量 Γ 与吸附质在溶液中的平衡浓度 c 有关。费罗因德里希（Freundlich）从吸附量和平衡浓度之间的关系曲线得到下面经验方程：

$$\Gamma = \frac{x}{m} = kc^n \tag{3.28}$$

式中，x 为吸附质的物质的量，mol；m 为吸附剂的质量，g；c 为吸附平衡时溶液的浓度，$mol \cdot L^{-1}$；k 和 n 都是经验常数，由温度、溶剂、吸附质与吸附剂的性质所决定，一般 $n<1$。将式(3.28) 取对数，可得：

$$\lg \Gamma = n\lg c + \lg k \tag{3.29}$$

根据此方程以 $\lg \Gamma$ 对 $\lg c$ 作图，可得一直线，由斜率和截距可求得 n 和 c。式(3.28)是经验方程，只适用于溶质浓度不太大和不太小的溶液。从公式上看，k 为 $c=1\mathrm{mol \cdot L^{-1}}$ 时的吸附量，但实际上此时式(3.28)可能已不适用。

朗格缪尔（Langmuir）从吸附过程的理论考虑，认为吸附量是单分子层吸附，即吸附剂一旦被吸附质占据之后，就不能再吸附。在吸附和脱附达成动态平衡时，推导出等温吸附方程式：

$$\Gamma = \Gamma_\infty \frac{cK}{1+cK} \tag{3.30}$$

式中，Γ_∞ 为饱和吸附量即每克吸附剂上被吸附质铺满单分子层时的吸附量，$\mathrm{mol \cdot g^{-1}}$；$\Gamma$ 为溶液在平衡浓度为 c（$\mathrm{mol \cdot L^{-1}}$）时的吸附量，$\mathrm{mol \cdot g^{-1}}$。将式(3.30)整理可得式(3.31)：

$$\frac{1}{\Gamma} = \frac{1}{\Gamma_\infty} + \frac{1}{\Gamma_\infty K} \times \frac{1}{c} \tag{3.31}$$

以 $1/\Gamma$ 对 $1/c$ 作图得一直线，由此直线的斜率和截距可求得 Γ_∞ 和常数 K，K 与吸附和脱附常数有关，与式(3.28)中的 k 意义不同。

根据 Γ_∞ 的数值，按照 Langmuir 单分子层吸附模型，并假定吸附质分子在吸附剂表面上是直立的，每个吸附质分子的截面积为 A_m，则吸附剂的比表面积 S（$\mathrm{m^2 \cdot g^{-1}}$）。可按下式计算：

$$S_0 = \Gamma_\infty N_A A_m \tag{3.32}$$

式中，N_A 为阿伏加德罗常数，$6.022 \times 10^{23} \mathrm{mol^{-1}}$；$A_m$ 为吸附质分子的横截面积，$\mathrm{m^2}$，每个醋酸分子横截面积为 $24.3 \times 10^{-20} \mathrm{m^2}$，根据水-空气界面上对于直链正构脂肪酸测定的结果而得。

活性炭对氮气的吸附是多分子层物理吸附，符合 BET 吸附等温式：

$$\frac{p/p}{V(1-p/p_s)} = \frac{1}{V_m C} + \frac{C-1}{V_m C} \times \frac{p}{p_s} \tag{3.33}$$

式中，p 为吸附平衡压力；p_s 为实验温度下液态氮气的饱和蒸气压；p/p_s 为相对压力；V 为在 p/p_s 时吸附量被换算成标准状态下的气体体积；V_m 为吸附质在吸附剂表面上形成单分子层时的吸附量，也换算成标准状态下气体的体积；C 为与吸附热有关的常数。

使用表面积分析仪在液氮控温为 -196℃ 时，测定不同相对压力 p/p_s 下（p/p_s 为 0.05~0.35）活性炭对氮气的吸附量 V 值，换算成标准状态下的气体体积，以 $\dfrac{p/p}{V(1-p/p_s)}$ 对 p/p_s 作图，得一直线，由其斜率和截距可得 V_m：

$$V_m = \frac{1}{\text{斜率} + \text{截距}} \tag{3.34}$$

活性炭的比表面积 S_0（$\mathrm{m^2 \cdot g^{-1}}$）为：

$$S_0 = \frac{V_m N_A \sigma}{22400 W} \tag{3.35}$$

式中，N_A 为阿伏加德罗常数，$\mathrm{mol^{-1}}$；W 为吸附剂的质量，g；σ 为一个吸附质分子的截面积，N_2 分子的截面积为 $16.2 \times 10^{-20} \mathrm{m^2}$。

三、仪器与试剂

表面积分析仪及其配套装置；恒温振荡机；磨口锥形瓶；移液管；滴定管。标准

NaOH 溶液；标准 HAc 溶液；活性炭。

四、实验步骤

1. 活性炭在醋酸水溶液中对醋酸的吸附

（1）取 6 个洗净的、带有磨口塞的锥形瓶，编号。按表 3.4 中给出的数据，配制各种不同浓度的醋酸水溶液。

（2）将 120℃下烘干的活性炭约 1g（准确称量至 0.001g），放入磨口锥形瓶中在恒温条件下振荡适当的时间（视温度而定，一般 0.5～2h，以吸附达到平衡为准）。振荡速度以活性炭可翻动为宜。

（3）使用颗粒状活性炭时，可用带有塞上棉花的橡皮管的移液管从磨口锥形瓶中吸取上部清液，按表 3.4 所列体积取样，用 $0.5\text{mol}\cdot\text{L}^{-1}$ 的标准 NaOH 溶液滴定。

（4）活性炭吸附醋酸是可逆吸附，因此使用过的活性炭可回收利用（用蒸馏水浸泡数次，烘干、抽气后即可）。

表 3.4 数据记录

编号	1	2	3	4	5	6
蒸馏水/mL	49	48	46	43	40	35
HAc 溶液/ $\text{mol}\cdot\text{L}^{-1}$	1	2	4	7	10	15
活性炭/g						
醋酸初始浓度 c_0/ $\text{mol}\cdot\text{L}^{-1}$						
取样量/mL						
滴定消耗 NaOH 溶液量/mL	40	40	20	20	20	20
醋酸平衡浓度 c/ $\text{mol}\cdot\text{L}^{-1}$						
吸附量 $\varGamma$/ $\text{mol}\cdot\text{g}^{-1}$						
$1/\varGamma$						
$1/c$						

2. 使用表面积分析仪测定活性炭对氮气的吸附

本实验使用 BECKMAN COULTER 公司的 SA3100 表面积分析仪，实验步骤如下。

（1）准确称取活性炭约 0.1g（样品比表面积不同，则取样量不同。按照仪器要求，样品比表面积大于 $30\text{m}^2\cdot\text{g}^{-1}$ 时，最佳样品量为 0.1～0.2g）。

（2）设置适当的实验参数，在 300℃下脱气 60min。

（3）脱气后再次准确称量活性炭的质量。

（4）按要求小心放置盛有液氮的保温瓶，仪器自动开始吸附实验。

首先，用氦气测量死体积。然后，在液氮（-196℃）环境中吸附 N_2，测定氮气相对压力为 0～0.2 时的吸附等温线。

仪器采用多分子吸附 BET 公式，自动分析、计算，给出标准状态下吸附数据、吸附等温线、$\dfrac{p/p}{V(1-p/p_s)}$ 对 p/p_s 的直线图及活性炭的比表面积。

说明：

① 液氮为-196℃液体，操作时应戴防护眼镜和手套，小心操作，避免溅到皮肤和眼睛。

② 实验过程中的操作要等待仪器提示后方可进行，以免损坏仪器或影响分析结果。

③ 具体操作步骤以 SA3100 表面积分析仪说明书及实验过程中仪器提示为准。

五、数据处理

1. 由醋酸平衡浓度 c 及初始浓度 c_0 数据按下式计算吸附量：

$$\Gamma = (c_0 - c)V/m$$

式中，V 为溶液的体积，L；m 为加入溶液中的吸附剂质量，g。

2. 作吸附量 Γ 对平衡浓度 c 的吸附等温线。
3. 计算 $\lg\Gamma$，$\lg c$，作 $\lg\Gamma$-$\lg c$ 图，由直线的斜率及截距求式(3.28) 中的常数 n 和 k。
4. 计算 $1/\Gamma$，$1/c$，作 $1/\Gamma$-$1/c$ 图，由直线的斜率及截距求 Γ_∞。
5. 由 Γ_∞ 根据式(3.35) 计算活性炭的比表面积。

六、注意事项

1. 操作过程中，应防止 HAc 的挥发，以免引起较大误差。
2. 本实验配制溶液用不含 CO_2 的蒸馏水。溶液配好摇匀后再放入活性炭。
3. 使用 SA3100 表面积分析仪要严格按操作规程进行操作。

七、思考题

1. 溶液吸附时，如何判断达到吸附平衡？
2. 为什么要测量死体积？
3. 用活性炭吸附溶液中的醋酸和活性炭吸附 N_2 这两种方法测得的比表面积大小有何不同？分析原因。

实验 3.12　油品燃烧热的测定

一、实验目的

1. 设计利用氧弹量热计测定油品物质的燃烧热。
2. 进一步掌握热化学实验的一般知识和测量技术。

二、实验原理

燃烧热是指 1mol（或单位质量）物质在氧气中完全燃烧时所释放出的热量。若燃烧在恒容下进行称为恒容燃烧热 (Q_V)，在恒压下进行称为恒压燃烧热 (Q_p)。由热力学第一定律可知，当燃烧在恒容下进行，体系不对外做功，恒容燃烧热 Q_V 等于系统的热力学能 U 的改变，即：$\Delta U = Q_V$。

燃烧热的测定在工业上常用于石油、煤、天然气、燃料油、液化石油气等的热值测量；在食品和生物学中用以计算营养成分的热值，据此指导营养滋补品合理配方的确定。本实验在前面有机物燃烧热测定基础上，仍然采用氧弹式量热计测定燃烧前后体系温度的改变量 ΔT，并通过放热总量等于吸热总量的原理求出柴油的燃烧热。

三、设计要求

1. 设计柴油燃烧热测定的实验步骤。
2. 计算出柴油的恒压燃烧热和恒容燃烧热。

四、仪器与试剂

SHR-15 氧弹式量热计；SWC-ⅡD 数字温度温差仪；氧气钢瓶；1000mL 容量瓶；电子天平；充氧器；尺子。柴油；引火丝。

五、实验提示

本实验采用自动点火，点火电流是恒定的。对于难以引燃的样品，为了保证能完全燃烧，可以在样品与燃烧丝之间缚一段棉纱线，以起助燃作用。棉纱线的参考燃烧热为 $-16.7\text{kJ}\cdot\text{g}^{-1}$。

用氧弹量热计测定液态物质燃烧热时，沸点高的液态物质可直接置于坩埚中，用引燃物引燃测定；对于沸点较低的物质，通常将其密封在已知燃烧热的胶管或塑料薄膜中，通过引燃物将其燃烧而测定。

六、思考题

1. 在燃烧热实验中，哪些是体系，哪些是环境？实验过程中有无热损耗？这些热损耗对实验结果有何影响，如何校正？
2. 在环境恒温式量热计中，为什么内筒水温要比环境温度低？低多少合适？
3. 燃烧热测定时采用什么方法确定量热计的热容值？测定操作中要注意什么？

实验 3.13　微溶盐浓度积的测定

一、实验目的

1. 根据所学可逆电池理论设计电池，测定氯化银的溶度积。
2. 学会 Ag-AgCl 电极的制备方法，理解制备过程的注意事项。

二、实验原理

电池电动势法是测定难溶盐溶度积的常用方法之一。测定氯化银的溶度积可以设计下列电池

$$\text{Ag(s)} | \text{AgCl(s)} | \text{KCl}(a_1) \| \text{AgNO}_3(a_2) | \text{AgCl(s)} | \text{Ag(s)}$$

Ag-AgCl 电极的电极电势可用下式表示

$$E(\text{Ag}^+|\text{AgCl}|\text{Ag}) = E^{\ominus}(\text{Ag}^+|\text{AgCl}|\text{Ag}) - \frac{2.303RT}{F}\lg a(\text{Cl}^-) \tag{3.36}$$

AgCl 的溶度积 K_{sp}：

$$K_{sp} = a(\text{Ag}^+)a(\text{Cl}^-) \tag{3.37}$$

将式(3.37)代入式(3.36)得

$$E(\text{Ag}^+|\text{AgCl}|\text{Ag}) = E^{\ominus}(\text{Ag}^+|\text{AgCl}|\text{Ag}) - \frac{2.303RT}{F}\lg K_{sp} + \frac{2.303RT}{F}\lg a(\text{Ag}^+) \tag{3.38}$$

电池的电动势为两电极电势之差：

$$E_{右} = E^{\ominus}(\text{Ag}^+|\text{AgCl}|\text{Ag}) - \frac{2.303RT}{F}\lg K_{sp} + \frac{2.303RT}{F}\lg a(\text{Ag}^+)$$

$$E_{左} = E^{\ominus}(\text{Ag}^+|\text{AgCl}|\text{Ag}) - \frac{2.303RT}{F}\lg a(\text{Cl}^-)$$

$$E = E_+ - E_- = \frac{2.303RT}{F}\lg K_{sp} + \frac{2.303RT}{F}\lg a(Ag^+)a(Cl^-)$$

整理后得

$$\lg K_{sp} = -\frac{EF}{2.303RT} + \lg a(Ag^+)a(Cl^-) \tag{3.39}$$

若已知银离子和氯离子的活度，测定了电池的电动势就能求出氯化银的溶度积。

三、设计要求

1. 设计电池并测定其电池电动势。
2. 根据测得的电池电动势计算氯化银的溶度积并与文献值比较。

四、仪器试剂

SDC 数字电位差综合测试仪（或 UJ-25 型电势差计）；检流计；标准电池；干电池。Ag-AgCl 电极、$AgNO_3$ 溶液（标定）、KCl 溶液（标定）、蒸馏水。

五、实验提示

1. 本实验所用试剂均为分析纯，溶液用重蒸馏水配制。

2. 本实验所用 Ag-AgCl 电极可采用电镀法进行制备。首先配制电镀所需的镀银液。分别将 $AgNO_3$（35~45g）、$K_2S_2O_5$（35~45g）、$Na_2S_2O_3$（200~250g）溶于 300mL 蒸馏水中。然后混合 $AgNO_3$ 和 $K_2S_2O_5$ 溶液，并不断搅拌使生成白色的胶状硫酸银沉淀，此后再加入 $K_2S_2O_5$ 溶液，并不断搅拌至白色沉淀全部溶解为止，加水稀释至 1000mL。新鲜配制的镀银溶液略显黄色并有少量混浊和沉淀，静置数日过滤得非常稳定的澄清镀银液。下面就是利用电镀法制备电极。将表面经过清洁处理的自制铂丝电极作阴极，把经砂纸打磨光洁的银丝电极作阳极，在镀银溶液中进行镀银。电流控制在 5mA 左右。40min 后即可在铂丝电极上镀上白色紧密的银层。将镀好的银电极用蒸馏水仔细冲洗，然后以此银电极为阳极，另选一铂丝或铂片电极作阴极，对 $0.1mol·L^{-1}$ HCl 溶液进行电解，电流仍控制在 5mA 左右，通电 20min 后就可在银电极表面形成 Ag-AgCl 镀层（呈紫褐色）。

3. 电池电动势的测量　参考前面基本实验 2.8 的操作步骤和要求。电池电动势的测定可将电池置于 25℃ 恒温槽中进行。测定时，电池电动势开始时可能不稳定，每隔一定时间测定一次，到测得稳定值为止。

六、思考题

1. 试分析有哪些因素影响实验结果？
2. 本实验会不会存在液接电势？简述消除液接电势的方法。

实验 3.14　极化曲线的测定

一、实验目的

掌握用恒电流（控制电流）法测定极化曲线的方法，了解极化曲线的物理意义及应用。

二、实验原理

在研究可逆电池的电动势和电池反应时，电极上几乎没有电流通过，每个电极或电池反应都是在无限接近于平衡（即准静态）状态下进行的，因此电极反应是可逆的。当有限的电流通过电池时，电极的平衡状态被破坏，此时电极反应处于不可逆状态，随着电极上电流密度的增加，电极反应的不可逆程度也随之增大。

在有电流通过电极时，由于电极反应的不可逆而使电极电势偏离平衡值的称作电极的极化，根据实验测出的数据来描述电流密度与电极电势之间关系的曲线称作极化曲线。通过极化曲线的测绘，可使我们对电极极化过程以及金属腐蚀与保护等加深认识和理解。

三、仪器与试剂

直流稳流电源（0～3A）或蓄电池 1 台；直流电流计（0-100-500mA）1 台；饱和甘汞电极 1 支；白金电极 1 支；碳钢（普通碳钢片）电极（电极面积为 10cm^2）2 片；UJ-25 型电位计（或数字电压表）1 套；氢气发生器或氮气钢瓶 1 套；停表 1 块；电解杯，小烧杯各 1 个；KNO$_3$ 盐桥 1 支；滑线电阻（2000Ω）2 个。

2mol·L^{-1}(NH$_4$)$_2$CO$_3$ 溶液；饱和 KCl 溶液；0.5mol·L^{-1} H$_2$SO$_4$ 溶液；金相砂纸；导线；石蜡；蓄电池。

四、实验步骤

1. 用金相砂纸将研究电极擦至镜面光亮放在丙酮中除去油污，用石蜡涂抹多余面积（若电极面积已按计划剪裁好，则不必再用石蜡涂抹），然后置于 0.5mol·L^{-1} H$_2$SO$_4$ 溶液中以研究电极作阴极，电流密度保持在 5mA·cm^{-2} 以下电解 10min 以除去氧化膜，最后用蒸馏水洗净备用（不用时可浸泡在有机溶剂如无水乙醇或丙酮中保存）。

2. 洗净器皿，于电解杯中倾入 2mol·L^{-1} (NH$_4$)$_2$CO$_3$ 溶液，按装置图 3.19 安装好测定阴极极化曲线的电极、参比电极及盐桥等。为了在测量电势时减小溶液欧姆电位降的影响，盐桥的尖嘴应尽量靠近研究电极的表面。连接线路时注意"＋"、"－"极不要接错。

3. 通电前，在电解液中通氢气（或氮气）5～10min（通氢气应在通风橱内进行），以除去溶液中的氧。

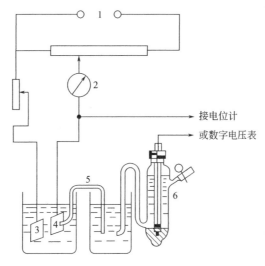

图 3.19　极化曲线测量装置示意图

1—电源及滑线电阻；2—电流计；3—辅助电极；4—研究电极；5—盐桥；6—参比电极（饱和甘汞电极）

4. 通电后,移动滑线电阻的滑动接头,控制电流密度为 $0.5mA \cdot cm^{-2}$,测定阴极和参比电极之间的电动势(测定时若检流计在平衡时波动较大,可将分流器旋钮指向"×0.1"或"×0.01"。开始时,每间隔 $0.5mA \cdot cm^{-2}$ 测定一次电动势);电流计到 50mA 以后,每间隔 $1mA \cdot cm^{-2}$ 测定一次电动势;电流计到 100mA 以后,每间隔 $2mA \cdot cm^{-2}$ 测定一次,直至电流计指针指到 480mA 时为止,然后重复一次。测定时要保持电流稳定,为了使测定的数据一致,每当电流计指针在指定的数值稳定 1~2min 后,立即测定电动势。

5. 阴极极化曲线测定后,将研究电极按步骤 1 进行电解处理;更换电解液,通氢气(或氮气)5~10min 以除氧,然后测定阳极极化曲线。测定时电流密度的间隔控制在 $0.3\sim0.5mA \cdot cm^{-2}$,当电流密度为 $10mA \cdot cm^{-2}$ 时,将电流密度间隔改为 $2.5mA \cdot cm^{-2}$,测到电流密度为 $30mA \cdot cm^{-2}$ 时为止。

做完实验后,将电极对调(或进行电解处理),重复测定一次。

记录实验测定时的温度和大气压。

五、数据处理

以电流密度(或电流密度的对数值)为纵坐标,电极电势(相对饱和甘汞电极)为横坐标绘出阴极和阳极极化曲线。

实验 3.15 不同反应体系化学振荡现象的初步研究

一、实验目的

1. 初步研究不同体系的化学振荡现象。
2. 进一步了解化学反应的特点及其产生化学现象的条件。
3. 初步认识化学反应体系在远离平衡态下,由于本身的非线性动力学机制而产生的宏观时空有序结构。

二、实验原理

化学振荡是一种周期性的化学现象。早在 17 世纪,波义耳就观察到磷放置在一瓶口松松塞住的烧瓶中时,会发生周期性的闪亮现象。1921 年,勃雷(W.C.Bray)在一次偶然的机会发现 H_2O_2 与 KIO_3 在硫酸稀溶液中反应时,释放出 O_2 的速率以及 I_2 的浓度会随时间周期变化。直到 1959 年,贝洛索夫首先观察到并随后为恰鲍廷斯基深入研究,丙二酸在溶有硫酸铈的酸性溶液中被溴酸钾氧化的反应,随后人们发现了一大批可呈现化学振荡现象的含溴酸盐的反应系统。人们统称这类反应为 B-Z 反应。

BrO_3^--丙二酸-Ce^{3+}-H_2SO_4 振荡反应体系(如实验 3.7),是目前研究最深入、完整的典型化学振荡体系。至今已发现一大批可呈现化学振荡现象的反应体系,有不少是在 B-Z 振荡反应的基础上变换而来。

1. 金属催化剂的替换

常见的有 Ce^{3+}/Ce^{4+},Mn^{3+}/Mn^{2+},$Fe[phen]_3^{3+}/Fe[phen]_3^{2+}$ 等。表 3.5 列出了 B-Z 振荡反应常用催化剂的一些特性。

表 3.5 B-Z 振荡反应常用催化剂的特性

电极反应	电极电势/V	颜色变化
$Ce^{3+} - e \longrightarrow Ce^{4+}$	1.44	黄色→无色
$Mn^{3+} + e \longrightarrow Mn^{2+}$	1.51	桃红色→无色
$Fe[phen]_3^{3+} + e \longrightarrow Fe[phen]_3^{2+}$	1.06	蓝色→红色

2. 无机酸的替换

经典 B-Z 反应对无机酸很敏感。改变 H_2SO_4 的浓度对振荡反应的诱导期、振荡周期都会有影响,一般要求其浓度大于 $0.4 mol \cdot L^{-1}$。其他离子的催化系统,对无机酸不特别敏感。例如 Mn^{3+}/Mn^{2+} 催化系统,可以在 H_3PO_4、HNO_3、$HClO_4$ 中进行化学振荡。但 HCl 一般情况下是不行的,因为 HCl 有严重干扰作用。

3. 有机物的替换

经典 B-Z 反应的有机物丙二酸可以由其他有机化合物来代替,这方面的研究已经相当广泛,报道较多的有机替代物见表 3.6。

表 3.6 常见有机替代物

有机替代物	催化剂	有机替代物	催化剂
丙酮	Ce^{3+}/Ce^{4+}, Mn^{3+}/Mn^{2+}	环己酮	Ce^{3+}/Ce^{4+}
柠檬酸	Ce^{3+}/Ce^{4+}	苯乙醇酸	Mn^{3+}/Mn^{2+}, $Fe[phen]_3^{3+}/Fe[phen]_3^{2+}$
没食子酸	Ce^{3+}/Ce^{4+}, Mn^{3+}/Mn^{2+}, $Fe[phen]_3^{3+}/Fe[phen]_3^{2+}$	二羟基苯甲酸	Ce^{3+}/Ce^{4+}, Mn^{3+}/Mn^{2+}, $Fe[phen]_3^{3+}/Fe[phen]_3^{2+}$

4. 其他物质的参与

例如:BrO_3^--木糖-丙酮-Mn^{2+}-H_2SO_4

BrO_3^--乳酸-丙酮-Mn^{2+}-H_2SO_4

BrO_3^--果汁-柠檬酸-Ce^{4+}-H_2SO_4

BrO_3^--氨基酸-丙酮-Mn^{2+}-H_2SO_4

BrO_3^--人血清-丙二酸-Mn^{2+}-H_2SO_4

5. 非溴离子控制的化学振荡反应

例如:IO_3^--丙二酸-Mn^{2+}-H_2O_2-H_2ClO_3

在通常的化学反应中,反应物和产物的浓度单调地发生变化,最终达到不随时间变化的平衡状态。然而,在化学振荡反应体系中,有些组分的浓度会忽高忽低,呈现周期性变化而远离平衡。在这类反应的研究中,动态跟踪不同实验条件下反应体系所呈现的周期性变化是实验研究的关键。采用的方法有电位法、分光光度法等。

化学振荡反应是基于振荡反应的参数如振荡的诱导期 $t_{诱}$、振荡周期 $t_{振}$、振荡寿命 t 及振幅 y 等受各种因素的影响而变化。这些因素有反应物的种类、浓度、反应温度、搅拌速度等。

三、实验要求

1. 查阅相应文献资料,设计并完成一个化学振荡反应实验。
2. 观察实验现象,研究物质的种类、反应物浓度、反应温度、搅拌速度对振荡系统参

数 $t_{诱}$、$t_{振}$ 及振幅 y 的影响。

四、仪器与试剂

参考实验 3.7。

五、实验提示

1. 一般化学振荡反应只能在很小的反应物浓度范围内发生，因此在确定了反应体系后，反应物浓度的选择很重要。表 3.7～表 3.10 为一些化学振荡反应系统反应物的初始浓度表，供参考。

表 3.7　BrO_3^--氨基酸-丙酮-Mn^{2+}-H_2SO_4 系统振荡反应

反应物	BrO_3^-	L-酪氨酸	丙酮	Mn^{2+}	H_2SO_4
初始浓度/mol·L^{-1}	0.040	0.042	0.054	0.016	1.00

表 3.8　BrO_3^--乳酸-丙酮-Mn^{2+}-H_2SO_4 系统振荡反应

反应物	BrO_3^-	乳酸	丙酮	Mn^{2+}	H_2SO_4
初始浓度/mol·L^{-1}	0.025	0.15	0.15	0.005	1.00

表 3.9　BrO_3^--果汁-柠檬酸-Ce^{4+}-H_2SO_4 系统振荡反应

反应物	BrO_3^-	橘子原汁	柠檬酸	Ce^{4+}	H_2SO_4
初始浓度/mol·L^{-1}	0.024	0.1mL	0.012	0.005	0.60

表 3.10　IO_3^--丙二酸-Mn^{2+}-H_2O_2-H_2ClO_3 系统振荡反应

反应物	IO_3^-	H_2O_2	丙二酸	Mn^{2+}	H_2ClO_3
初始浓度/mol·L^{-1}	0.024	0.1mL	0.012	0.005	0.60

注：1. 所列反应物中还要加入 0.6mol·L^{-1} 的淀粉作为指示剂。
2. 注意反应物的加入顺序，一般催化剂最后加入。

六、思考题

1. 影响诱导期、周期及振荡寿命的主要原因有哪些？
2. 试举例介绍一些化学振荡反应方面的应用实例。

第 4 章 实验技术与仪器

4.1 大气压力计

4.1.1 大气压力计的测量原理

大气压力是用与大气压力相平衡的汞柱高度来测量的。国际上规定在纬度 45°处海平面及温度为 0℃时 760mm 高的汞柱所平衡的大气压力为标准大气压。大气压力计就是按此规定计量的。

大气压力计的测量原理如图 4.1 所示。

设管内汞柱上方为真空，当汞柱与大气压力平衡时表现出的压力为：

$$p_a = \rho g H \tag{4.1}$$

式中，p_a 为大气压力，Pa；ρ 为汞的密度，$kg \cdot m^{-3}$；g 为重力加速度，$m \cdot s^{-2}$；H 为汞柱高，m。

汞的密度 ρ 与测量时的温度有关。重力加速度随纬度及海拔高度而异。因此，用大气压力计测量时，除了进行仪器误差的校正外，在精密测量中还必须进行温度、纬度和海拔高度方面的校正。

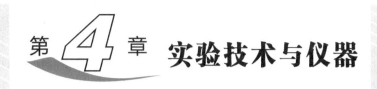

图 4.1 大气压力计的测量原理

4.1.2 大气压力计的结构

大气压力计的式样有多种，实验室常用的是福丁（Fortin）式气压计（图 4.2）。

大气压力计外部是一黄铜管，上有刻度标尺 E。管内装一玻璃管，其上端封闭，下端开口并插入水银槽 A 中。玻璃管内有水银柱，管内水银柱上面为真空。黄铜管上半段前后开有长方形窗孔，以便观察玻璃管内水银面的位置。在窗孔部位安装一游标尺 F，旋转螺丝 G 可调节游标尺的上下位置。铜管中部附有温度计 H。水银槽 A 底部有羚羊皮，羚羊皮下面有螺丝 C 支持，转动 C 可以升降水银槽内水银面的高低，水银槽上部有一象牙针 D，针尖的位置是刻度标尺 E 的零点。

4.1.3 大气压力计的使用方法

先旋转底部螺丝 C，升高槽 A 内的水银面，使水银面与象牙针 D 的尖端恰好接触。再旋转气压计中部螺丝 G，将游标尺 F 升起至比玻璃管内水银面稍高的位置。然后使游标尺 F 下降，直到游标尺的前后下缘恰与水银柱的凸面相切。上述调节中注意眼睛应与水银面在同一水平面上。按照游标尺的下缘零线所对的刻度标尺上的位置读出大气压力测

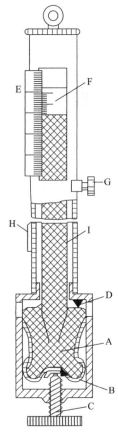

图 4.2 大气压力计
A—水银槽；B—羚羊皮膜；C—底部螺丝；
D—象牙针；E—标尺；F—游标尺；
G—中部螺丝；H—温度计；
I—玻璃管（上端封死）

量值的标准的准确数字部分，而可疑数字部分用游标尺来决定。即从游标尺上找出一根正好与标尺 E 上某一刻度相重合的刻度线，此刻度线的读数为大气压力测量值的最后一位数字。现使用的大气压力计标尺有用高度（厘米）及压力（毫巴）标度的。如图 4.3 所示标尺为用高度标度的，其中（a）刻度应读为 76.03cm，（b）刻度应读为 76.12cm。大气压力计的读数应记录四位有效数字。同时应记下大气压力计上的温度，准备校正用。

大气压力计必须垂直安装。如果偏离垂直位置 1°，则对 76.00cm 来说，会造成 0.01cm 误差。

4.1.4 大气压力计测量值的校正

（1）温度的校正　温度会影响水银的密度及黄铜标尺的长度。针对这两个因素，可得到下列校正公式。

$$H_0 = H_t - \frac{H_t(\beta-\alpha)t}{1+\beta t} \quad (4.2)$$

式中，H_0 为将水银柱校正到 0℃ 时的高度；H_t 为水银柱在 t℃ 时的高度；α 为黄铜的线膨胀系数，1.84×10^{-5}℃$^{-1}$；β 为汞的体膨胀系数，1.818×10^{-4}℃$^{-1}$；t 为读数时的温度，℃。

为了使用方便，将已知数据代入得：

$$\Delta_t H = H_0 - H_t = -\frac{1.634 \times 10^{-4} t}{1+1.818 \times 10^{-4} t} H_t \quad (4.3)$$

应该指出，精密测量中气压计上的温度计也要校正。如果此温度计读数偏差 1°，则对于 76.00cm 的测量值会引入 0.012cm 的误差。

经过温度校正的大气压力测量值单位为厘米（0℃）汞柱高。若实验中要将大气压力与水银压差计的汞柱高相加、减求绝压时，则应将水银压差计所测汞柱高也换算为 0℃ 汞柱高后再加减，求得绝压的单位也是厘米（0℃）汞柱高。

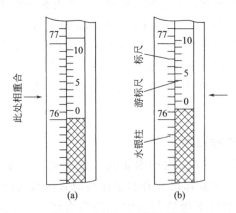

图 4.3　标尺、游标示意图

(2)重力加速度 g 的校正 在纬度 $45°$ 的海平面重力加速度是 $9.80665 \mathrm{m \cdot s^{-2}}$。当纬度及海拔高度改变时，重力加速度有所改变。因此各地重力加速度下测得的汞柱高应换算成上述标准重力加速度下的汞柱高。当纬度为 ϕ（°），海拔高度为 h（m）时，对已校正到 $0℃$ 的汞柱高度再作如下校正：

$$H_g = H_0(1-2.65\times10^{-3}\cos2\phi)(1-3.14\times10^{-7}h)$$
$$\approx H_0(1-2.65\times10^{-3}\cos2\phi-3.14\times10^{-7}h)$$

由此式可看出纬度校正值为：

$$\Delta_\phi H = -2.65\times10^{-3}H_0\cos2\phi \tag{4.4}$$

海拔高度校正值为：

$$\Delta_h H = -3.14\times10^{-7}hH_0 \tag{4.5}$$

(3)仪器的修正值 这是由于气压计构造上的缺陷或长期使用后水银中溶解微量空气渗入真空部分所引起的。当与标准气压计相比较后，可求得此项修正值。这项修正值常附于仪器的检定证书中。

(4)高度差的校正 当气压计水银槽 A 中水银面与实验进行所在位置存在高度差时，实验环境大气压力与气压计测量值之间存在差别。通常地球表面的 10cm 高空气柱大致相当 0.9mm 汞柱的压力。即每升高 10m，大气压力减少 0.9mm 汞柱高。

经上述诸项校正后，大气压的实际值为：

$$H = H(读数) + H(仪器误差) + \Delta_\phi H + \Delta_t H + \Delta_h H \tag{4.6}$$

校正举例：

北京地区纬度近似为 $39°56'$，海拔为 100m，若气压计上温度计读数为 $25℃$ 时，气压计观测值为 76.53cm，已知仪器误差为 $+0.01$cm，试计算校正后的大气压力。

温度校正：

$$\Delta_t H = -\frac{1.634\times10^{-4}\times25}{1+1.818\times10^{-4}\times25}\times76.53 = -0.311\mathrm{cm}$$

纬度校正

$$\Delta_\phi H = -2.65\times10^{-3}\times\cos(2\times39.93)\times76.53 = -0.036\mathrm{cm}$$

海拔校正

$$\Delta_h H = -3.14\times10^{-7}\times100\times76.53 = -0.002$$

实际大气压 $H = 76.53+0.01-0.311-0.036-0.002 = 76.19$cm

$$p_a = 76.19\times1333.2 = 1.0158\times10^5 \mathrm{Pa}$$

故大气压力应为 $1.0158\times10^5 \mathrm{Pa}$

为了方便，对于一个已安装好的气压计可把仪器修正、纬度及海拔修正合并成一个修正值。在要求不高的场合下，也可以只做温度修正。

有的大气压力计以毫巴标度，$1\mathrm{bar}=1000\mathrm{mbar}=10^5\mathrm{P}$。上述各校正公式中 H_t 值可直接代入毫巴读数，求得各修正值的单位也是毫巴。

4.2 温度控制系统

恒温槽中常用的温度控制系统有：①温度控制仪；②继电器、电接触温度计组合系统。

4.2.1 温度控制仪

（1）工作原理　温度指示控制仪（图 4.4 和图 4.5）由直流电压电桥比较器、控温执行继电器及感温元件等部分组成。当感温探头热敏电阻感受的实际温度低于控制选择温度时，继电器触电闭合，加热线路连通，使控温介质开始加热（此时加热指示灯亮）。当感温探头热敏电阻感受温度与控制选择温度相同或略高时，继电器触电断开，停止加热（此时恒温指示灯亮），重复上述过程达到自动控温目的。

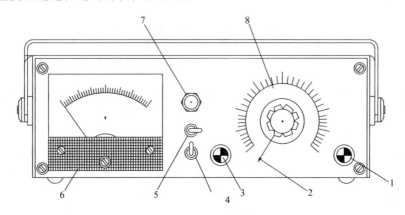

图 4.4　WMZK-01 温度指示控制仪面板图

1—控温继电器断开指示（红）；2—起始圆点；3—控温继电器接通指示（白）；4—电源开关；
5—测温满程转换开关；6—测温指示电表；7—测温满程调节；8—温控选择盘

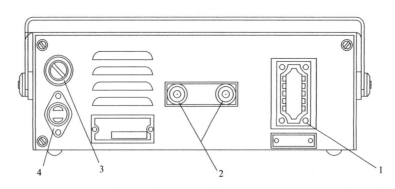

图 4.5　WMZK-01 温度指示控制仪器后板面图

1—感温探头插座；2—控温继电器输出接柱；3—电源保险丝；4—仪器电源插座

测温系统与支流电桥直接送入运算放大器，驱动温度指示电表，可直接观察升降温度值，而得到被测温度的结果。

（2）操作步骤

① 按要求连接线路，并将感温探头置于恒温介质中，打开电源开关。

② 将控温调节旋钮调至欲恒定之温度（如 30℃），此时温控仪器将自动控温。因长期使用，控温调节旋钮零点易发生漂移，导致控温指示不准确，调节方法如下：

将控温调节旋钮调至略高于室温（应比实际要控制的温度低），记录此值为 T_1，此时温控仪命令加热器加热（加热指示灯亮），介质温度上升，当介质温度上升至等于指定控制温度 T_1 时，加热器停止加热（此时恒温指示灯亮），准确读出恒温介质中的水银精密温度计上的读数（T_2），则控温调节旋钮的误差为：

$$\Delta T = T_2 - T_1$$

如 ΔT 接近零，则控温调节旋钮零点无漂移，如 ΔT 不为零，可根据其正负校对控温调节旋钮的零点。

③ 温控仪上的温度指示可能产生零点误差应在开启电源之前用螺丝刀调节，使指针指示在起始温度线上，通常介质中的温度应该从介质中悬挂的精密水银温度计读出。

4.2.2 继电器、电接触温度计组合系统

（1）晶体管继电器 温度控制器包括感温元件及控制器两部分。本书实验 2.1 中介绍的接触温度计是常用的灵敏度较高的一种感温元件。热敏电阻、热点偶等也可用作感温元件，控制器有多种形式，下面介绍配合接触温度计使用的晶体管继电器（图 4.6）。

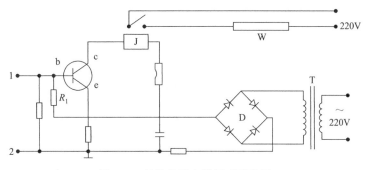

图 4.6 晶体管继电器原理线路图

图 4.6 中右侧为电源部分，左侧为晶体管继电器部分。三极管的基极电流由 200kΩ 的电阻 R_1 限制在 120μA 左右，使集电极的电流略大于继电器 J 的工作电流。当接触温度计中汞柱与钨丝未接触时，1、2 点间断路，三极管的集电极电流使继电器工作，电加热 W 通电加热，恒温槽温度上升。当温度达到控制温度，汞柱与钨丝接触，1、2 点间短路，基极电流为零，集电极电流很小，继电器将衔铁放开，电加热器停止加热。

71 型晶体管继电器接线如图 4.7 所示。2、3 是继电器电源线接头，7、8 为感温元件引线接头，4、5、6 三个接线柱中有两个接控制负载。接 4、5，则 7、8 间断路时，负载通电工作（此为恒温槽的接触温度计与电加热器所要求的）；接 5、6，则 7、8 间为通路时，负载通电工作（此为控制低温时，以制冷机为负载的情况）。使用时应注意继电器的功率与负载要相适应，对于电抗性负载，继电器的负荷能力为纯电阻时的 60%。

（2）电子管继电器 图 4.8 为电子管继电器的工作原理图。图中 L 为电子管，一般为

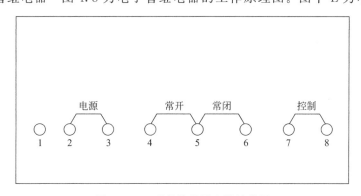

图 4.7 71 型晶体管继电器接线图

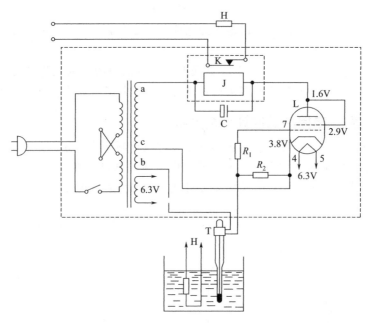

图 4.8　电子管继电器的工作原理图

L—6P1 电子管；T—接触温度计；H—加热器；J—继电器线圈；K—衔铁；R_1，R_2—电阻；C—滤波电容

栅流较小的 6P1 或 6N1 电子管，把板极和帘栅接在一起再接继电器的线圈 J，然后接到变压器一端，阴极和变压器的抽头 e 相接，这时栅极未与变压器 b 点相接（接触温度计断开）。所以，6P1 可看成一个二极管。当变压器 a、e 两端的交流电压加在电子管的板极和阴极之间时，则电子管中就有一个半波脉动直流通过，经电容 C 滤波后成稳定的直流，通过继电器的线圈 J 产生一磁场，吸引弹簧片，使电键 K 闭合，因此使加热器 H 通电。水温升高，接触温度计 T 中水银上升，与钨丝相接，使变压器 b 端与电子管栅极相通。通常在电子管中只有变压器 a 端为正半周时才有可能导电，所以当栅极成自由栅时，6P1 是有电流通过的。可是当栅极与 b 相接时，由于这时 b 端的电位比 c 点更负，则电子管中电流变小或截止。继电器断开，H 不加热，恒温槽水温下降，同时接触温度计水银面下降，并与钨丝断开（相当于 b 与 7 断开），则电子管栅极负电压消失，6P1 又导电，J 中又产生磁场，加热器又被导通，如此反复进行，就使水温恒定在某一比较稳定的温度。

4.3　饱和蒸气压减压装置

4.3.1　减压装置简介

减压装置（图 4.9）与数字压力表、实验装置、气泵配套使用，用于饱和蒸气压的测定。

4.3.2　使用说明

（1）安装　将压力罐阀 1 及两只三通接口用橡胶管或塑料管分别与气泵、压力表、实验装置连接即可。安装时应注意连接管插入接口的接入深度≥15mm，否则会影响气密性。

（2）操作　将压力罐阀（以下简称阀 1）、系统调压阀（以下简称阀 2）打开，微调阀

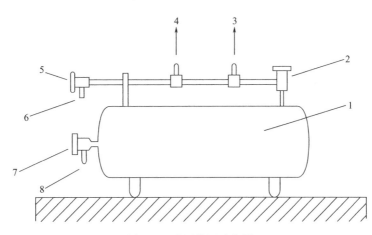

图 4.9 减压装置示意图
1—压力罐；2—阀2；3—实验装置；4—压力表；5—阀3；6—大气；7—阀1；8—气泵

（以下简称阀3）关闭，（方向——顺时针关闭，逆时针开启，阀1、2、3相同）启动气泵加压或抽气，压力表所显示的数字即为压力罐中的压力。首次使用或长期使用，重新起用时应先做密封性实验。操作过程：关闭阀1，停止气泵工作，检查阀2是否开启、阀3是否完全关闭。观察压力表，显示数字下降值在标准范围（每分钟压力下降≤0.1kPa即为正常），气密性良好，超出标准范围需查找原因，进行维修。

经气密性试验，并证明性能良好后，可进入实验操作：打开阀1、启动气泵，当压力达到实验所需的压力（应略高于实验所需最高压力值）时，首先关闭阀1，停止气泵工作，关闭阀2，用阀3将压力调正到实际所需数值，即可开始实验，实验过程中需调压力值时，先打开阀2，将压力罐中的压力输入实验系统内，当压力表上显示的压力值略高于实验所需压力值时即关闭阀2，用阀3调整至所需压力值。采用本法可获取实验过程所需的不同压力值。保证整个实验的圆满完成。实验完毕，应打开阀2、阀3释放压力使系统处于常压下备用。

4.4 阿贝折光仪

折射率是物质的重要物理性质之一，可借助它了解物质的纯度、浓度及其结构。在实验室中常用阿贝折光仪来测量物质的折射率。测量液态物质折射率时，它具有试样用量少、操作方便、读数精确度高等优点。

4.4.1 阿贝折光仪的构造原理

光从一种介质进入另一种介质时，其传播速度发生变化，传播方向也发生变化，即发生光的折射现象，在定温下单色光在两种介质的界面上，入射角 i 的正弦与折射角 r 的正弦之比等于该单色光在介质1和介质2中传播速度 u_1、u_2 之比（见图4.10）。即：

$$\frac{\sin i}{\sin r} = \frac{u_1}{u_2} = n_{12}$$

n_{12} 称为折射率，当介质1为真空（或空气）时，它为物质的特性常数。对一定波长的光在一定温度、压力下，对于给定介质其折射率为定值。

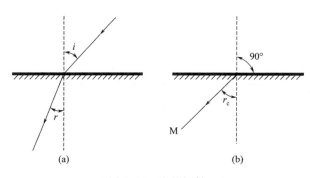

图 4.10 光的折射

当 $n_{12} > 1$ 时，由上式可知，i 角必大于 r 角。这时光由第一种介质进入第二种介质时将折向法线。在一定温度、压力下，光的入射角 i 增大至 90° 时，折射角达到极大值 r_c，$r_c < 90°$，r_c 称为临界折射角。因此，如图 4.10 所示入射角为 0°～90° 的全部光线，进入第二种介质后都应落在临界折射角 r_c 之内。若在 M 处置一目镜，则从目镜中见到的是半明半暗视野。由折射率的定义可知，当固定一种介质时，临界折射角 r_c 的大小和折射率（代表第二种介质的性质）有简单的函数关系。阿贝折光仪正是根据此原理设计制造的。

阿贝折光仪的外形如图 4.11 所示。仪器的主要部件为两个带恒温夹套的直角棱镜 5、6，两棱镜合拢时（图上为打开状态）中间有一狭缝，可由加样孔 9 加入被测液体，填满狭缝，使得两棱镜间夹了一层被测液体。由反射镜 7 将白光射入辅助棱镜 6 中。

折光仪中光的行程见图 4.12。辅助棱镜 P_1 的斜面是磨毛的面，而测量棱镜 P_r 的斜面是磨光的平面。当白光通过辅助棱镜时，可透过磨毛面漫射，得到各方向的光透过缝隙中的液层，从各个方向进入测量棱镜 P_r。如前所述，从各方向进入棱镜 P_r 的光线均发生折射，而它们的折射角均落在临界折射角 r_c 之内。在 r_c 之内的光线通过阿密西棱镜消除色散。光再经透镜聚焦后，在目镜中显出半明的像。当转动棱镜使目镜中见到的明暗分界线调到目镜中的十字线交叉点时，从读数目镜 2（见图 4.12）即可在标尺 S 上读出液体折射率。

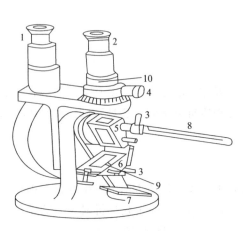

图 4.11 阿贝折光仪

1—读数目镜；2—测量目镜；3—循环恒温水接头；
4—消色补偿器；5—测量棱镜；6—辅助棱镜；
7—平面反射镜；8—温度计；9—加样孔；
10—校正螺丝

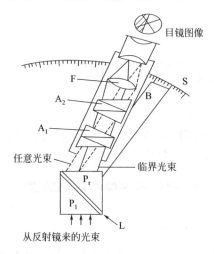

图 4.12 光的行程

P_r—测量棱镜；P_1—辅助棱镜；A_1，A_2—阿密西棱镜；
F—聚焦透镜；B—转动镜；S—标尺；L—被测液体层

实验测定时所用的光源常是白光。白光是各种不同波长的光混合而成的。由于波长不同的光在相同介质中的传播速度不同，在折射时会发生色散作用。因此，当用白光作入射光时，目镜中的明暗分界线不清楚。为了消除色散，在仪器中装有消色补偿器——阿密西棱镜。测定时转动消色补偿器4（图4.11），即可消除色散而得到清楚的明暗分界线。这时测得的液体折射率和应用钠光D线（波长为589.3mm）所得折射率相同。

在折光仪测量目镜筒上，有一个供校准用的螺丝。在校正折光仪时，先在棱镜夹套中通入恒温水循环。在狭缝内加入标准液体，并将刻度尺上读数调节为该温度下标准液体的已知折射率。此时若从测量目镜中看到明暗界线与交叉点不重合，则可转动校正螺丝，使分界线移动到交叉点。

折光仪的另一种校正方法为用仪器所附的标准玻璃校正。用一滴α-溴代萘液体将标准玻璃块光面粘在测量棱镜5的光面上，玻璃块侧面的光面向上。不要合上辅助棱镜6，打开测量棱镜5背面小窗的金属盖，使光线从小窗射入。然后将刻度尺上读数调至标准玻璃块上标出的折射率数值，用校正螺丝将测量目镜中的明暗界线调至交叉点。

4.4.2 使用方法

（1）将折光仪放在明亮处（但应避免阳光直射）。装上温度计。将恒温槽的恒温水通入棱镜夹套。在室温变化不大时，也可不接恒温水，在室温下测定。

（2）转动辅助棱镜上的螺栓，向下打开棱镜，向前转动折光仪主体部分，使测量棱镜斜面水平向上。在两棱镜的斜面上滴丙酮数滴，用擦镜纸沾干，将棱镜洗净。

（3）如测定黏度较大的不易挥发液体的折射率，可将1~2滴样品滴在测量棱镜的斜面上，合上辅助棱镜，进行测定。测定易挥发性液体样品的折射率时，则应先合上辅助棱镜，由侧面加样品孔加入样品，使被测液体充满两棱镜间的狭缝，进行测定。

（4）将折光仪主体部分转到适当位置，调节反射镜，使测量目镜中视野明亮清晰。调仪器侧面刻度盘下的大旋钮，转动棱镜，直至在测量目镜中看到明暗界线。

（5）调节消色补偿器的旋钮，消除目镜视野中的色散光，得到清楚的明暗界线。

（6）仔细调节大旋钮，转动棱镜，使测量目镜中明暗界线恰好与交叉点重合。

（7）从读数目镜中读出折射率数值。用同样方法，测定同一液体折射率的误差不应超过0.0002。同时记下温度计上的温度。

（8）打开棱镜（仍使测量棱镜斜面向上），用擦镜纸沾去样品，用丙酮洗净棱镜，使丙酮挥发干后，合上棱镜，待下次使用。

（9）用毕折光仪，应将棱镜面洗净。关闭恒温水，将金属夹套内的水倒净。卸下温度计，将折光仪放入仪器箱内保存。折光仪是精密、贵重的光学仪器，使用时要细心操作。

4.4.3 注意事项

（1）棱镜是折光仪的关键部件，要特别注意保护。开闭棱镜时不得用力过大。清洗棱镜时必须用擦镜纸沾干，不得往返地擦镜面。不得用其他纸、布或手擦镜面。

（2）用滴管从加样品孔滴入样品，不得将滴管插入孔内，防止滴管破损。如有破损，并有碎玻璃进入孔内时，应立即打开棱镜，用丙酮冲洗，用擦镜纸轻轻地沾去碎玻璃防止碎玻璃划坏或压坏棱镜镜面。

（3）不得测量酸性、碱性和其他腐蚀性液体。

(4) 保持仪器各部件（如目镜、反射镜等）的清洁，防止磨损。

(5) 折光仪切勿被阳光暴晒。

4.5 精密数字压力计

4.5.1 技术指标及说明

DP-AF 精密数字压力计（图 4.13），适用于负压测量及饱和蒸气压测定实验，可替代 U 形水银压力计测量压力差。测量范围：$-100\sim 0$ kPa；分辨率：0.01kPa；电源：$220V\pm 10\%$，50Hz；使用环境温度：$-10\sim 50$℃；湿度：$\leqslant 85$℃；传递介质：除氟化物以外的各种气体。

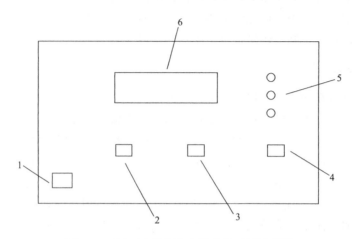

图 4.13　DP-AF 精密数字压力计面板示意图
1—电源开关；2—校正；3—采零；4—单位选择；5—单位指示；6—压力显示

4.5.2 使用方法

(1) 准备

① 用 $\phi 4.5\sim 6$mm 内径的真空橡胶管将仪表后盖板的压力传感器与被测系统连接。

② 将仪表后盖板的电源插座与 220V 电源连接。

③ 打开电源开关，此时仪表处于初始状态，预热 2min。

(2) 操作步骤

① 预压及气密性检查　缓慢加压到满量程，观察数字压力表值变化情况，若 1min 内显示值稳定，说明传感器及其被测系统无泄漏。确认无泄漏后，泄压至零，并在全量程反复预压 2~3 次，方可正式测量。

② 复零　泄压至零，使压力传感器通大气，按下复零键，以消除仪表系统的零点漂移。此时显示屏显示"0000"。

注意：尽管仪表作了精细的零点补偿，但因传感器本身固有的漂移（如时漂）是无法处理的，因此，每次测试前都必须进行复零操作，以保证所测压力值的准确度。

③ 测试　仪表复零后接通被测系统，此时仪表显示被测系统的压力值。

④ 关机　先将被测系统泄压后，再关掉电源开关。

4.6 数字式贝克曼温度计

4.6.1 特点

数字式贝克曼温度计（SWC-Ⅱ，见图 4.14），其测量精度高、范围宽、操作简便，还可与微机直接联结完成温度、温差的检测，实现自动化控制。

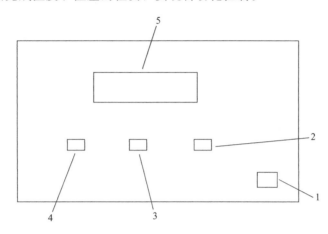

图 4.14 SWC-Ⅱ数字式贝克曼温度计面板示意图
1—电源开关；2—基温选择；3—温度与温差测量选择；4—测量与保持选择；5—数字显示屏

SWC-Ⅱ型数字贝克曼温度计的技术指标：测量范围 $-50 \sim 150$℃；分辨率 0.01℃；温差测量范围 ± 19.999℃；电源 220V$\pm$10V，50Hz；环境温度 $0 \sim 40$℃，相对湿度$\leqslant$85%。

4.6.2 使用方法

（1）测量前的准备

① 将仪器后面板的电源线接入 220V 电源插座。

② 检查感温插头编号（应与仪器后盖的编号相符）并将其和后盖的"Rt"端子对应连接紧（槽口对准）。

③ 将探头插入被测物中深度应大于 50mm，打开电源开关。

（2）温度测量

① 将面板"温度-温差"按钮置于"温度"位置（抬起位），显示器显示数字并在末尾显示"℃"，表明仪器处于温度测量状态。

② 将面板"测量-保持"按钮置于"测量"位置（抬起位）。

③ 按被测物的实际温度调节"基温选择"，使读数的绝对值尽可能的小。实际温度可用本仪器测量，记录 T_1 数值。

【例 4.1】物体实际温度为 15℃，则将"基温选择"置于 20℃ 位置，此时显示器显示 5.000℃ 左右。

④ 显示器动态显示的数字即为相对于 T_1 的温度变化量 ΔT。

【例 4.2】当 $T_1=5.835$℃ 时（基温位置不变），若显示器显示 6.325℃，则 $\Delta T=(6.325-5.835)$℃$=0.490$℃。

（3）保持功能的作用 当温度和温差变化太快无法读数时，可将面板"测量-保持"按钮置于"保持"位置（按下位）。读数完毕后转换到"测量"位置，跟踪测量。

4.7 DDS-11A 型电导率仪

（1）仪器的结构　DDS-11A 仪器的外形如图 4.15 所示。

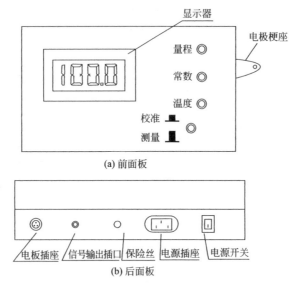

图 4.15　DDS-11A 仪器的外形示意图

① "量程"选择开关　分 2、20、200、2000μS·cm^{-1} 及 2×10^4 μS·cm^{-1} 五个测量挡。
② "常数"调节器　按所用电极的常数，调节使之显示相应的数值。
③ "温度"补偿器　使用前将其调节在液温的实际温度值的刻度线。
④ "校准/测量"按钮开关　按下时为"校准"，再按一下（开关向上弹起）则为"测量"。

（2）使用方法
① 接通电源，让仪器预热约 10min。
② 将温度按钮置于被测液的实际温度相应的位置上。当"温度"旋钮置于 25℃ 位置时，则无补偿作用。
③ 将电极浸入被测液体，电极插头插入插座。
④ 按下"校准/测量"开关，使其置于"校准"状态，调节"常数"旋钮，使仪器显示所用电极的常数标称值。如：电极常数为 0.95，调节旋钮显示 950，常数为 1.10，调节"常数"旋钮使显示 1100。
⑤ 按下"校准/测量"开关，使其处于"测量"状态（这时开关向上弹起），将"量程"开关置于合适的量程挡，待仪器示值稳定后，该显示数值即为被测液体在 25℃ 时的电导值。在测量中，若显示屏首位为 1，后三位数字熄灭，表明被测值超出量程范围，应置于高一挡量程来测量。若读数很小，则就置于低一挡量程，以提高精度。

4.8 补偿法原理及 UJ-25 型高电势直流电位差计

测定可逆电池电动势常用补偿法（又称对消法）。补偿法原理图如图 4.16 所示。

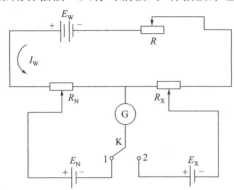

图 4.16 补偿法测电动势原理示意图

E_W—工作电源；E_N—标准电池；E_X—待测电池；R—调节电阻；
R_X—待测电池电动势补偿电阻；K—转换电键；
R_N—标准电池电动势补偿电阻；I_W—工作电流；G—检流计

电位差计是按照补偿法测量原理而设计的一种平衡式电学测量装置，能直接给出待测电池的电动势值。电位差计由三个回路组成：工作电流回路、标准回路和测量回路。

(1) 工作电流回路 也叫电源回路。从工作电源正极开始，经电阻 R_N、R_X，再经工作电流调节电阻 R，回到工作电源负极。其作用是借助于调节 R 使在补偿电阻上产生一定的电位降。

(2) 标准回路 从标准电池的正极开始（当换向开关 K 扳向"1"一方时），经电阻 R_N，再经检流计 G 回到标准电池负极。其作用是校准工作电流以标定补偿电阻上的电位降。通过调节 R 使 G 中电流为零，此时 R_N 产生的电位降与标准电池的电动势 E_N 相对消，也就是说大小相等而方向相反。校准后的工作电流，I_W 为某一定值，即 $I_W = E_N/R_N$。

(3) 测量回路 从待测电池的正极开始（当换向开关 K 扳向"2"一方时），经检流计 G 再经电阻 R_X 回到待测电池负极。在保证校准后的工作电流 I_W 不变，即固定 R 的条件下，调节电阻 R_X，使得 G 中电流为零。此时 R_X 产生的电位降与待测电池的电动势 E_X 相对消，即 $E_X = I_W R_X$，则 $E_X = (E_N/R_N) R_X$。

为使工作电流等于给定值，需要利用标准电池。标准电池的电动势是已知的。例如某标准电池在实验温度下的电动势为 1.01863V，要求工作电流 I_W 为 0.100000mA，我们将 R 的阻值调为 10186.3Ω，将换向开关 K 扳到 2 一边，如果检流计光点不动，则工作电流恰为 0.100000mA。如果检流计光点偏转，则调节可变电阻 R，直到光点不偏转。这一步骤称为标定工作电流。

用补偿法原理测电动势所用仪器称为电位差计。电位差计根据其测量范围和精度，有多种型号。UJ-25 型高电势直流电位差计是实验室用的精密度高、测量范围大的直流电位差计。它可用于测量直流电位差，配用标准电阻时也可测量直流电流强度和电阻，以及检验功率表。

4.8.1 原理线路简介

在工作电池的"＋"、"－"钮间为电位差计的工作电流回路。其电流由粗到细地用 4 个串联在电路里的可变电阻 R（见图 4.17）调节至 0.0001A。工作电流回路其余部分由下列各部件组成。

① 第Ⅰ测量十进盘，由 18 个 1000Ω 的电阻组成。其中第 5 个电阻由 1 个 999Ω 电阻及温度补偿 B 的十进盘、10 个 0.1Ω 电阻串联而成。第 16 个由 180Ω、8101Ω 电阻及温度补偿 A 的十进盘、10 个 1Ω 电阻串联而成。此 18 个 1000Ω 电阻中每个两端电位差为 0.1V，共 1.8V。

此十进盘中第 6～15 个 1000Ω 电阻（共 10 个）、180Ω 电阻及温度补偿 A、B 电阻的一部分构成标准电池补偿电阻，供标定工作电流用。

② 第Ⅱ测量十进盘由 11 个 100Ω 电阻组成。其中有 10 个电阻两端电位差为 0.01V。另一个与第Ⅴ、Ⅵ测量盘并联的电阻两端电位差为 0.009V。

③ 第Ⅲ测量十进盘由 10 个 10Ω 电阻组成。其中每个电阻两端电位差 0.001V。

④ 第Ⅳ测量十进盘由 10 个 1Ω 电阻组成。其中每个电阻两端电位差为 0.0001V。

⑤ 第Ⅴ测量十进盘由 10 个 1Ω 电阻组成，第Ⅵ测量十进盘由 10 个 0.1Ω 电阻组成。此两测量盘与一个 889Ω 电阻串联，并始终与第Ⅱ测量盘中某一个 100Ω 电阻并联。此并联电路总电阻为 90Ω，总电流等于工作电流，为 0.0001A，故其两端总电位差为 0.009V。而第Ⅴ测量盘上每个电阻两端电位降为 1×10^5 V，第Ⅵ测量盘上每个电阻两端电位降为 1×10^6 V。

上述 6 个测量盘中第Ⅲ、Ⅳ、Ⅴ三个测量盘分别有替代盘，此替代盘与十进盘共有 20 个相同电阻，而在旋转旋钮时始终只有 10 个电阻串联进工作电流回路，故只起 10 个电阻的作用。

综上所述，UJ-25 型电位差计中工作电流回路中测量部分总电阻为 19200Ω。工作电流为 0.001A（在第Ⅴ、Ⅵ测量盘中为 1×10^5 A），总电位差为 1.9200V。测量上限为 1.911110V。

标准电池（接"标准"钮）、待测电池（接"未知"钮）、检流计（接"电计"钮）及"粗"、"细"按钮组成补偿线路（见图 4.17）。"粗"、"细"按钮并联。"粗"按钮上串联了 500kΩ 的高电阻，以保护检流计。"短路"按钮连"电计"两钮，使检流计本身短路用。

电位计上有两组"未知"钮并联，可接两个待测电池（见图 4.18）。标定电流或分别测量待测电池电动势可用转换开关选择。

电位计上有两组接工作电池的按钮，共 3 个（见图 4.18），可选用两种电动势不同的直流电源中的一种。当工作电池电动势为 1.95～2.2V（接一节铅蓄电池）时工作电流回路总内阻在 19200～23553Ω 范围内（包括可变电阻 R 及测量部分）。当工作电池电动势为 2.9～3.3V（接两节甲电池）时工作电流回路中串联了一个 9650Ω 电阻，起降压作用。此时工作电流回路总电阻在 28850～33203Ω 范围内。

4.8.2 测量电动势时电位差计的使用方法

在使用电位差计前先将转换开关放在"断"位置，将左下方的 3 个按钮全部松开。按图 4.18 的位置接检流计、工作电池、标准电池及待溅电池。接各电池时注意正、负极位置。

(1) 调整工作电流　首先将标准电池温度补偿 A、B 调到标准电池的电动势值（此值由标准电池电动势与温度关系式，按实验时的室温算出）。然后将转换开关 K 放在 N 位置上，按下"粗"按钮，使高电阻 r 接入测量电路，避免检流计和标准电池中电流过大而受损伤。检流计光点发生偏转，立即松开按钮。根据光点偏转方向由粗至细地调可变电阻 R。反复多次按"粗"按钮及调节，直至检流计光点偏转不明显。再按"细"按钮调可变电阻 r，直至检流计光点不偏转为止。此时工作电流为 0.0001A。调节中，松开"粗"、"细"按钮时若检流计光点摆动不停，可按"短路"按钮，使光点较快停在零点。

(2) 测量待测电池电动势　将转换开关 K 放在"×1"或"×2"位置上。首先按"粗"按钮，调节第Ⅰ～Ⅵ测量盘旋钮，使检流计光点偏转不明显。再按"细"按钮，调节Ⅳ～Ⅵ测量盘旋钮，使检流计光点不偏转为止。从 6 个读数窗孔内读出待测电池电动势的各位数字。

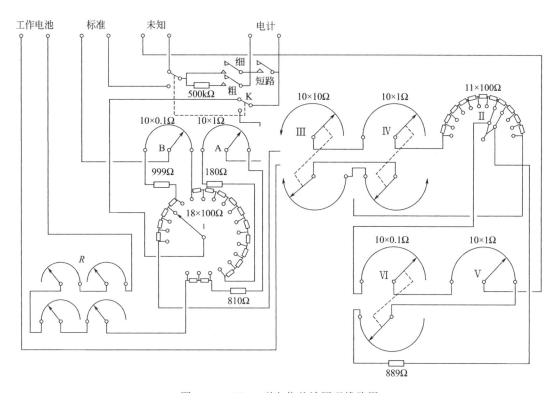

图 4.17 UJ-25 型电位差计原理线路图

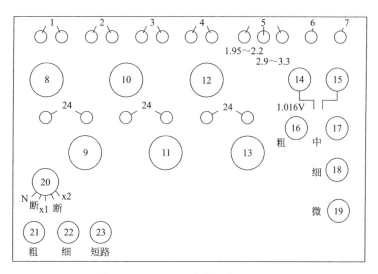

图 4.18 UJ-25 型电位差计板面布置

1—检流计接线柱；2—标准电池接线柱；3，4—待测电池接线柱；5—工作电池接线柱；6—泄漏屏蔽；7—静电屏蔽；
8，9，10，11，12，13—第Ⅰ、Ⅱ、Ⅲ、Ⅳ、Ⅴ、Ⅵ测量盘旋钮；14，15—标准电池温度补偿 A，B；
16，17，18，19—调工作电流可变电组 R；20—转换开关 K；21—粗调按钮；
22—细调按钮；23—短路按钮；24—读数窗孔

注意：在测量前最好预先估计待测电动势大小，将各测量盘调至此值，减小检流计光点偏转，缩短测量时间。在按按钮时应尽量缩短时间，减少通过标准电池或待测电池的电量，减少极化带来的偏差。

以上方法仅能测量1.911110V以下的电位差。若需测量1.911110V以上的电位差，可配用适当的分压箱以提高电位差计的测量上限。具体使用方法可见该仪器说明书。

4.9 旋光仪的构造原理及使用方法

旋光仪是研究溶液旋光性的仪器，用来测定平面偏振光通过具有旋光性物质的旋光度的大小和方向，从而定量测定旋光物质的浓度；确定某些有机物分子的立体结构。

4.9.1 构造原理

一般光源发出的光，其光波在与传播方向垂直的一切可能方向上振动，这种光称为自然光，或称为非偏振光，而只在一个固定方向有振动的光称为偏振光。

当一束自然光投射到各向异性的晶体（例如方解石，即$CaCO_3$晶体）中时，产生双折射。折射光线只在与传播方向垂直的一个可能方向上振动，因此可分解为两束互相垂直的平面偏振光，从而获得了单一的平面偏振光。

旋光仪的主要部件尼科耳棱镜就是根据这一原理设计的。尼科耳棱镜是由两个方解石直角棱镜所组成（如图4.19所示）。棱镜两锐角为68°和22°；两棱镜直角边用加拿大树胶黏合起来（图中AD）。当自然光以一定的入射角投射到棱镜时，双折射产生的O光线在第一块直角棱镜与树胶交界面上全反射，为棱镜框子上涂黑的表面所吸收。双折射产生的光线则透过树胶层及第二个棱镜而射出。从而在尼科耳棱镜的出射方向上获得了一束单一的平面偏振光。这个尼科耳棱镜称为起偏镜，它是被用来发生偏振光的。

目前多数应用某些晶体的二色性来制成偏振光。它是在一个薄片的表面上涂一薄层（约0.1mm）二色性很强的物质的细微晶体（如硫酸碘-金鸡纳霜或硫酸金鸡纳碱等），能够吸收全部寻常光线，从而得到偏振光。

偏振光振动平面在空间轴向角度位置的测量也是借助于一块尼科耳棱镜，这里称为检偏镜。它是由偏振片固定在两保护玻璃之间，并随刻度盘同轴转动。当一束光经过起偏镜后光沿OA方向振动，如图4.20所示。也就是可以允许在这一方向上振动的光通过此平面。OB为检偏镜的透射面，只允许在这一方向上振动的光通过。两透射面的夹角为θ。振幅为E的OA方向的平面偏振光可以分解为振幅分量分别为$E\cos\theta$和$E\sin\theta$的两互相垂直的平面偏振光，并且只有$E\cos\theta$分量（与OB相重）可以透过检偏镜，而$E\sin\theta$分量不能透过。当$\theta=0°$时，$E\cos\theta=E$，此时透过检偏镜的光最强；当$\theta=90°$时，$E\cos\theta=E$，此时没有光透过检偏镜，光最弱。如以I表示透过检偏镜的光强；I_0表示透过起偏镜入射的光强。当θ角在0°～90°之间变化时，则有以下关系：

$$I = I_0 \cos^2\theta$$

旋光仪就是通过透光强弱明暗来测定其旋光度。在起偏镜与检偏镜之间如放置被测物质时，由于被测物质旋光作用，原来由起偏镜出来的偏振光转过一个角度，因而检偏镜也相应转过一个角度，只有这样才能使透过的光强与原来相同。

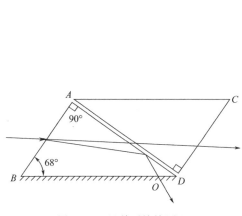

图 4.19　尼科耳棱镜原理

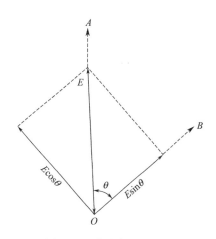

图 4.20　旋光仪测量原理

由于实际观测上肉眼对视野场明暗程度的感觉是不甚灵敏的，为了精确地确定旋转角，常采取比较的办法，即三分视场（也有二分视场）的方法，在起偏片后的中部装一狭长的石英片，其宽度约为视野的 1/3，由于石英片具有旋光性，从石英片中透过的那一部分偏振光被旋转了一个角度 φ，因为 $\angle AOB$ 为 $90°$，$\angle COB$ 不等于 $90°$，所以在望远镜中透过石英片的那部分稍暗，两旁是黑暗的。旋光仪三分视场如图 4.21（a）所示。当 $\angle POB$ 为 $90°$ 时，因 $\cos^2(\angle AOB)$ 等于 $\cos^2(\angle COB)$，视野中三个区内的明暗相等，此时三分视场消失，视场均黑，如图 4.21（c）所示。当 $\angle POB$ 为 $180°$ 时，整个视场均匀明亮，如图 4.21（d）所示。人的视觉对明暗均匀与不均匀有较大敏感。我们在实验中采用图中（c）的视野，而不采用图（d）视野，因这时视场显得特别明亮，不易辨别三个视场的消失。

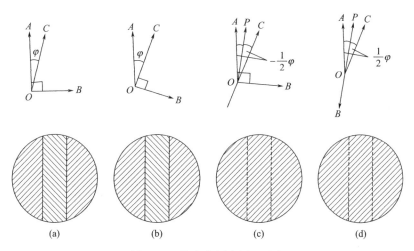

图 4.21　旋光仪测定原理示意

4.9.2　使用方法

使用旋光仪时，先接通电源，开启开关，约 5min 以后，钠光灯发光正常后才可开始工作。将样品管充满蒸馏水，盖好玻璃片，施好压紧螺帽，使样品管不泄漏。亦不可旋得过紧，以免引起玻璃片应力，影响读数准确性。样品管中若有小气泡，应将其赶至样品管的扩大部分。将样品管擦干净，放入旋光仪。根据需要，接通恒温水，循环恒温。先调目镜焦

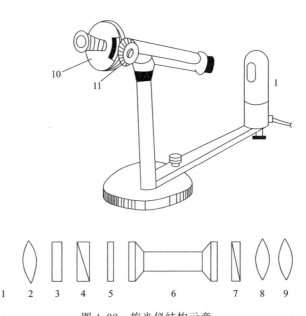

图 4.22 旋光仪结构示意

1—钠灯光；2—透镜；3—滤光片；4—起偏镜；5—石英片；6—旋光管；7—检偏镜；
8，9—望远镜透镜；10—刻度圆盘；11—传动轮

距，使视野清晰。再调节刻度盘手轮，找到等暗面（零度视场），读取刻度值，作为仪器零点。在样品管中更换为待测溶液，用上述同样方法，测出其刻度值，将此值减去零点值即得样品溶液在此实验条件下的旋光度。旋光仪结构如图 4.22 所示。

为提高读数精度，仪器采用光学游标跳线对准读数装置，左、右有两个读数窗口。读数时先找到游标 0 刻度线对应的刻度盘读数。再找出游标刻度盘刻度线对齐的位置，读游标读数。左、右两个窗口分别读数，再取平均值作为测量结果。旋光仪连续使用不得超过 4h。

4.10　WLS-2 可调式恒流电源

WLS-2 可调式恒流电源（图 4.23）采用电流、电压双组取样并经单片机处理后显示，其读数直观、准确。由于仪器采用了负载四线制取样，从而消除了负载电阻对电压显示的影响。电路的电压限幅，使得实验更安全、可靠。该仪器输出电流为 $0\sim2A$，输入电压为 $0\sim15V$，电流分辨率为 1mA，电压分辨率为 0.01V。

4.10.1　使用方法

（1）将负载按颜色插入并略旋紧（地线按需要与正或负连接，也可不连接）。

（2）接通电源，打开电源开关。

（3）将两线夹短路，调节粗、细调整旋钮，使电流显示所需读数。

（4）将负载夹与负载按正负极性相连，此时仪器显示通过负载的电流及两端电压。

4.10.2　注意事项

（1）本仪器仅适用于约 220V 电网。

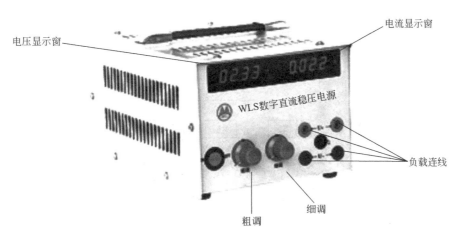

图 4.23　WLS-2 可调式恒流电源

(2) 使用中，也可将两正接线端子或负接线端子同时连接，但此时仪器电压指示为本仪器正、负端子间的电压，而非负载两端的电压。

(3) 恒流供电时，电流下降，电压为恒定值，则说明负载电阻过大，不能进入正常恒流供电。

(4) 本仪器应在短路情况下先调整好电流后，再接通负载，以免电流过大损坏负载。

(5) 切勿用金属等导电物接触后盖散热器上的调整管壳体，以免仪器内部短路而损坏仪器。

4.11　SWC-ⅡD 精密数字温度温差仪

在物理化学实验中，对体系的温差进行精确测量（如燃烧热和中和热的测定）时，以往都是使用水银贝克曼温度计。这种水银玻璃仪器虽然原理简单、形象直观，但使用时易破损，且不能实现自动化控制，特别是在使用前的调节比较麻烦，近年来逐渐被电子贝克曼温度计所取代。电子贝克曼温度计的测量探头（热电偶）通常采用对温度极为敏感的热敏电阻，它是由金属氧化物半导体材料制成的，其电阻与温度的关系为 $R=Ae^{-B/t}$（R 为电阻；t 为温度，℃；A、B 为与材料有关的常数）。通过温度的变化，转换成电性能变化，测量电性能变化便可以测量出温度的变化。

SWC-ⅡD 精密数字温度温差仪属于电子贝克曼温度计的一种，其操作面板如图 4.24 所示。该仪器可同时测量体系的温度和温差，且具有精度高、测量范围宽和操作简单等优点，此外还设有可调报时、读数保持、基温自动选择及超量程显示等功能，并配备 RS-232C 串行口，可以实现温度温差检测与控制自动化。

4.11.1　使用方法

(1) 将传感器探头插入被测物中，深度大于 5cm，打开电源开关。开机后，仪器即显示传感器探头所测物的温度。

(2) 温差测量

① 基温选择：仪器根据被测物温度，自动选择合适的基温，其基温选择的标准如表 4.1 所示。

图 4.24 SWC-ⅡD 精密数字温度温差仪

表 4.1 基温选择标准

温度 T	基温 T_0	温度 T	基温 T_0
$T<-10℃$	$-20℃$	$50℃<T<70℃$	$60℃$
$-10℃<T<10℃$	$0℃$	$70℃<T<90℃$	$80℃$
$10℃<T<30℃$	$20℃$	$90℃<T<110℃$	$100℃$
$30℃<T<50℃$	$40℃$	$110℃<T<130℃$	$120℃$

注：基温下 T_0 不一定为绝对准确值，其为标准温度的近似值。

② 温差显示：面板温差显示部分即为被测物实际温度 T 与基温 T_0 的差值。

(3) "采零"键的应用　当温差显示值稳定时，可按"采零"键，使温度显示为"0.000"，仪器将此时的被测物温度 T 当作 0，若被测物温度变化时，则温差显示的即为温度的变化值。

(4) "锁定"键的应用　在一个实验过程中，仪器"采零"后，当被测物温度变化过大时，仪器的基温会自动选择，这样，温差的显示值将不能正确反映温度的变化值，所以在实验开始后，按"采零"键后，再按"锁定"键，则仪器将不会改变基温。此时"采零"键也不起作用，直至重新开机。

(5) "测定/保持"键的使用　当温度和温差的变化太快无法读数时，可将面板"测定/保持"按钮置于"保持"位置，读数完毕后再转换到"测定"位置，跟踪测量。

(6) 定时读数　按面板"▲"或"▼"键，调至所需的报时间隔。调整完后，定时显示倒计时，当一个计数周期完毕后，蜂鸣器鸣叫，且读数保持约 5s，此时可观察和记录数据。若不想报警，只需将"定时"读数置于"0"即可。

4.11.2　注意事项

(1) 在测量过程中，"锁定"键需慎用，一旦按"锁定"键后，基温自动选择和"采零"将不起作用，直至重新开机。

(2) 当仪器杂乱无章或显示"OUL"时，表明仪器温差测量已超量程，应检查被测物的温度或传感器是否连接好，且重新"采零"。

(3) 仪器数字不变时，可检查仪器是否处于"保持"状态。

附录　物理化学常用数据表

附表1　国际相对原子质量

原子序	名称	符号	相对原子质量	原子序	名称	符号	相对原子质量
1	氢	H	1.0079	31	镓	Ga	69.723
2	氦	He	4.00260	32	锗	Ge	72.59
3	锂	Li	6.941	33	砷	As	74.9216
4	铍	Be	9.01218	34	硒	Se	78.96
5	硼	B	10.81	35	溴	Br	79.904
6	碳	C	12.011	36	氪	Kr	83.80
7	氮	N	14.0067	37	铷	Rb	85.4678
8	氧	O	15.9994	38	锶	Sr	87.62
9	氟	F	18.99840	39	钇	Y	88.9059
10	氖	Ne	20.179	40	锆	Zr	91.22
11	钠	Na	22.98977	41	铌	Nb	92.9064
12	镁	Mg	24.305	42	钼	Mo	95.94
13	铝	Al	26.98154	43	锝	Tc	98.9062
14	硅	Si	28.0855	44	钌	Ru	101.07
15	磷	P	30.97376	45	铑	Rh	102.9055
16	硫	S	32.06	46	钯	Pd	106.4
17	氯	Cl	35.453	47	银	Ag	107.868
18	氩	Ar	39.948	48	镉	Cd	112.41
19	钾	K	39.0983	49	铟	In	114.82
20	钙	Ca	40.08	50	锡	Sn	118.69
21	钪	Sc	44.9559	51	锑	Sb	121.75
22	钛	Ti	47.90	52	碲	Te	127.60
23	钒	V	50.9415	53	碘	I	126.9045
24	铬	Cr	51.996	54	氙	Xe	131.30
25	锰	Mn	54.9380	55	铯	Cs	132.9054
26	铁	Fe	55.847	56	钡	Ba	137.33
27	钴	Co	58.9332	57	镧	La	138.9055
28	镍	Ni	58.69	58	铈	Ce	140.12
29	铜	Cu	63.546	59	镨	Pr	140.9077
30	锌	Zn	65.39	60	钕	Nd	144.24

续表

原子序	名称	符号	相对原子质量	原子序	名称	符号	相对原子质量
61	钷	Pm	[147]	85	砹	At	[210]
62	钐	Sm	150.4	86	氡	Rn	[222]
63	铕	Eu	151.96	87	钫	Fr	[223]
64	钆	Gd	157.25	88	镭	Ra	226.0254
65	铽	Tb	158.9254	89	锕	Ac	227.2078
66	镝	Dy	162.50	90	钍	Th	232.0381
67	钬	Ho	164.9304	91	镤	Pa	231.0359
68	铒	Er	167.26	92	铀	U	238.029
69	铥	Tm	168.9342	93	镎	Np	237.0482
70	镱	Yb	173.04	94	钚	Pu	[244]
71	镥	Lu	174.963	95	镅	Am	[243]
72	铪	Hf	178.49	96	锔	Cm	[247]
73	钽	Ta	180.9479	97	锫	Bk	[247]
74	钨	W	183.85	98	锎	Cf	[251]
75	铼	Re	186.207	99	锿	Es	[254]
76	锇	Os	190.2	100	镄	Fm	[257]
77	铱	Ir	192.22	101	钔	Md	[258]
78	铂	Pt	195.09	102	锘	No	[259]
79	金	Au	196.9665	103	铹	Lr	[260]
80	汞	Hg	200.59	104			
81	铊	Tl	204.37	105			
82	铅	Pb	207.2	106			
83	铋	Bi	208.9804	107			
84	钋	Po	[209]				

注：引自 Robex C. Weast，"Handbook of Chemistry and Physics" 58th cd. B-1.

附表 2　水的饱和蒸气压

$T/℃$	p^*/kPa	$T/℃$	p^*/kPa	$T/℃$	p^*/kPa	$T/℃$	p^*/kPa
0	0.6105	13	1.497	26	3.361	39	6.991
1	0.6567	14	1.599	27	3.565	40	7.375
2	0.7058	15	1.705	28	3.780	45	9.583
3	0.7579	16	1.817	29	4.005	50	12.33
4	0.8134	17	1.937	30	4.242	55	15.73
5	0.8723	18	2.064	31	4.493	60	19.92
6	0.9350	19	2.197	32	4.754	65	25.00
7	1.002	20	2.338	33	5.030	70	31.16
8	1.073	21	2.486	34	5.320	75	38.54
9	1.148	22	2.644	35	5.624	80	47.34
10	1.228	23	2.809	36	5.941	85	57.81
11	1.312	24	2.984	37	5.275	90	70.096
12	1.403	25	3.168	38	6.625	95	84.513

注：引自"International Critical Tables of Numerical Data, Physics, Chemistry and Technology" Ⅲ-211.

附表 3 水的表面张力

单位：$N \cdot m^{-1}$

$T/℃$	$\sigma \times 10^3$	$T/℃$	$\sigma \times 10^3$	$T/℃$	$\sigma \times 10^3$	$T/℃$	$\sigma \times 10^3$
0	75.64	17	73.19	26	71.82	60	66.18
5	74.92	18	73.05	27	71.66	70	64.42
10	74.22	19	72.90	28	71.50	80	62.11
11	74.07	20	72.75	29	71.35	90	60.75
12	73.93	21	72.59	30	71.18	100	58.85
13	73.78	22	72.44	35	70.38	110	56.89
14	73.54	23	72.28	40	69.56	120	54.89
15	73.49	24	72.13	45	68.14	130	52.84
16	73.34	25	71.97	50	67.91		

注：引自《苏联化学手册》，科学出版社，1958，Ⅰ-977.

附表 4 水的黏度

单位：$Pa \cdot s$

$T/℃$	$\eta \times 10^3$	$T/℃$	$\eta \times 10^3$	$T/℃$	$\eta \times 10^3$	$T/℃$	$\eta \times 10^3$
0	1.787	10	1.307	20	1.002	30	0.7975
1	1.728	11	1.271	21	0.9779	35	0.7194
2	1.671	12	1.235	22	0.9548	40	0.6529
3	1.618	13	1.202	23	0.9325	45	0.5960
4	1.567	14	1.109	24	0.9111	50	0.5468
5	1.510	15	1.139	25	0.8904	60	0.4665
6	1.472	16	1.109	26	0.8705	70	0.4042
7	1.428	17	1.081	27	0.8513	80	0.3547
8	1.386	18	1.053	28	0.8327	90	0.3147
9	1.346	19	1.027	29	0.8148	100	0.2818

注：引自"Handbook of Chemistry and Physics"58th. F-51.

附表 5 甘油水溶液黏度

单位：$Pa \cdot s$

甘油(质量分数)/%	黏度 $\eta \times 10^3$			甘油(质量分数)/%	黏度 $\eta \times 10^3$		
	20℃	25℃	30℃		20℃	25℃	30℃
55	7.997	6.532	5.494	21	1.829	1.592	1.403
50	6.050	5.041	4.247	20	1.769	1.542	1.360
45	4.715	3.967	3.380	19	1.715	1.495	1.320
40	3.750	3.181	2.731	18	1.664	1.453	1.281
35	3.040	2.600	2.249	17	1.614	1.411	1.244
30	2.501	2.157	1.876	16	1.565	1.370	1.207
29	2.410	2.082	1.812	15	1.517	1.331	1.174
28	2.324	2.008	1.752	14	1.473	1.292	1.143
27	2.242	1.934	1.695	13	1.431	1.256	1.112
26	2.167	1.870	1.641	12	1.390	1.221	1.082
25	2.095	1.814	1.590	11	1.350	1.186	1.052
24	0.025	1.754	1.541	10	1.311	1.153	1.024
23	1.957	1.699	1.494	5	1.143	1.010	0.900
22	1.892	1.644	1.447	0	1.005	0.893	0.800

注：引自《苏联化学手册》，科学出版社，1958 年，Ⅲ-451.

附表 6　几种液体的饱和蒸气压

饱和蒸气压与温度的关系为：　　$\ln p^* = A + BT^{-1} + C\ln T + DT$

式中　p^*——饱和蒸气压，Pa；

　　　T——热力学温度，K。

$A、B、C、D$ 常数如下表所列：

物质	分子式	A	B	C	D	温度范围
水	H_2O	72.5295	−7101.70	−7.36514	4.40634×10^{-3}	0～100℃
乙醇	C_2H_5OH	−77.7155	−2785.53	18.7684	−0.0365627	0～90℃
正丙醇	C_3H_7OH	−29.6750	−4660.63	10.5744	−0.0236391	0～100℃
苯[①]	C_6H_6	7.49288	−4119.06	3.37990	−0.0117236	0～100℃
环己烷	C_6H_{12}	240.467	−10510.98	−37.0264	0.0510316	0～100℃
正庚烷	C_7H_{16}	285.127	−12243.73	−44.5109	0.0612472	0～100℃

[①] 此原始数据来自《苏联化学手册》，科学出版社，1958年，Ⅰ-799.

注：原始数据来自"International Critical Tables of Numerical Data, Physics, Chemistry and Technology" Ⅲ-217.

附表 7　醋酸的标准电离平衡常数

$T/℃$	$K_a^{\ominus} \times 10^5$	$T/℃$	$K_a^{\ominus} \times 10^5$	$T/℃$	$K_a^{\ominus} \times 10^5$
0	1.657	20	1.753	40	1.703
5	1.700	25	1.754	45	1.670
10	1.729	30	1.750	50	1.633
15	1.745	35	1.728		

注：引自"Handbook of Chemistry and Physics" 58th. D-152.

附表 8　不同温度下氯化钾的溶解热

溶解条件：1mol 氯化钾溶于 200mol 水。

温度/℃	溶解热/kJ	温度/℃	溶解热/kJ	温度/℃	溶解热/kJ
10	19.98	17	18.77	24	17.70
11	19.79	18	18.60	25	17.56
12	19.62	19	18.45	26	17.41
13	19.45	20	18.30	27	17.27
14	19.28	21	18.15	28	17.14
15	19.10	22	18.00	29	17.00
16	18.93	23	17.85		

附表 9　KCl水溶液的电导率　　　　　单位：$S \cdot m^{-1}$

温度 $T/℃$	KCl溶液浓度 c/mol·L^{-1}			
	1	0.1	0.02	0.01
10	8.319	0.933	0.1994	0.1020
15	9.252	1.048	0.2242	0.1147
20	10.207	1.167	0.2501	0.1278
21	10.400	1.191	0.2553	0.1305
22	10.554	1.215	0.2606	0.1332
23	10.789	1.239	0.2659	0.1359
24	10.984	1.264	0.2712	0.1386
25	11.180	1.288	0.2765	0.1413
26	11.377	1.313	0.2819	0.1441
27	11.574	1.337	0.2873	0.1468
28		1.362	0.2927	0.1496
29		1.387	0.2981	0.1524

续表

温度 $T/°C$	KCl溶液浓度 $c/mol \cdot L^{-1}$			
	1	0.1	0.02	0.01
30		1.412	0.3036	0.1552
31		1.437	0.3091	0.1581
32		1.462	0.3146	0.1609
33		1.488	0.3201	0.1638
34		1.513	0.3256	0.1667
35		1.539	0.3312	

附表10 压力单位换算

压力单位	Pa	$kgf \cdot cm^{-2}$	$dyn \cdot cm^{-2}$	$lbf \cdot in^{-2}$	atm	bar	mmHg
Pa	1	1.019716×10^{-5}	10	1.450342×10^{-4}	9.86923×10^{-6}	1×10^{-5}	7.5006×10^{-3}
$kgf \cdot cm^{-2}$	9.80665×10^4	1	9.80665×10^5	14.223343	0.967841	0.980665	735.559
$dyn \cdot cm^{-2}$	0.1	1.019716×10^{-6}	1	1.450342×10^{-5}	9.86923×10^{-7}	1×10^{-6}	7.5006×10^{-4}
$lbf \cdot in^{-2}$	6.89476×10^3	7.030396×10^{-2}	6.89476×10^4	1	6.80460×10^{-2}	6.89476×10^{-2}	51.7149
atm	1.01325×10^5	1.03323	1.01325×10^6	14.6960	1	1.01325	760.0
bar	1×10^5	1.019716	1×10^6	14.5038	6.986923	1	750.062
mmHg	133.3224	1.35951×10^{-3}	133.3224	1.93368×10^{-2}	1.3157895×10^{-3}	1.333221×10^{-3}	1

注:1. $\rho_{Hg}=13.5931g \cdot cm^{-3}$, $g=9.80665m \cdot s^{-2}$。0℃;1mmHg=1Torr=1/760atm。
2. 引自印永嘉主编.物理化学简明手册.北京:高等教育出版社,1998:668。

附表11 能量单位换算

能量单位	cm^{-1}	J	cal	eV
cm^{-1}	1	1.98648×10^{-23}	4.74778×10^{-24}	1.239852×10^{-4}
J	5.03404×10^{22}	1	0.239006	6.241461×10^{18}
cal	2.10624×10^{23}	4.184	1	2.611425×10^{19}
eV	8.065479×10^3	1.602189×10^{-19}	3.829326×10^{-20}	1

注:引自印永嘉主编.物理化学简明手册.北京:高等教育出版社,1998:668。

附表12 气瓶颜色标志一览

序号	充装气体名称	化学式	瓶色	字样	字色	色环
1	乙炔	CH≡CH	白	乙炔不可近火	大红	
2	氢	H_2	淡绿	氢	大红	$p=20MPa$,淡黄色单环 $p=30MPa$,淡黄色双环
3	氧	O_2	淡(酞)蓝	氧	黑	
4	氮	N_2	黑	氮	淡黄	$p=20MPa$,白色单环 $p=30MPa$,白色双环
5	空气		黑	空气	白	
6	二氧化碳	CO_2	铝白	液化二氧化碳	黑	$p=20MPa$,黑色单环
7	氨	NH_3	淡黄	液氨	黑	
8	氯	Cl_2	深绿	液氯	白	
9	氟	F_2	白	氟	黑	
10	一氧化氮	NO	白	一氧化氮	黑	
11	二氧化氮	NO_2	白	液化二氧化氮	黑	
12	碳酰氯	$COCl_2$	白	液化光气	黑	
13	砷化氢	AsH_3	白	液化砷化氢	大红	

参考文献

[1] 邝生鲁,刘常坤,向建敏等. 物理化学实验. 武汉:武汉工业大学出版社,1996:140-145.
[2] 胡广林,张雪梅,徐宝荣. 分析化学实验. 北京:化学工业出版社,2010.
[3] 章永年,梁治齐. 液体洗涤剂. 第2版. 北京:中国轻工业出版社,2000:642-645.
[4] 金丽萍,邬时清,陈大勇. 物理化学实验. 上海:华东理工大学出版社,2005.
[5] 郑传明,吕桂琴. 物理化学实验. 北京:北京理工大学出版社,2005.
[6] 武汉大学. 物理化学实验. 武汉:武汉大学出版社,2004.
[7] 傅献彩,沈文霞,姚天扬. 物理化学(下册). 第5版. 北京:高等教育出版社,2006.
[8] 天津大学. 物理化学(下册). 第4版. 北京:高等教育出版社,2001.
[9] Shoemaker D P, Garland C W, Nibler J W. Experiments in Physical Chemistry. 5th edn. New York:MnGraw-Hill Book Company, 1989.
[10] Weast R C. CRC Handbook of Chemistry and Physics. Boca Raton, Florida:CRC Press Inc, 1985-1986.
[11] 朱京,陈卫,金贤德,蔡显鄂. 液体燃烧热和苯共振能的测定. 化学通报,1984(3):50.
[12] Benton Brooks Owen. J Am Chem Soc, 1938, 60:2229.
[13] 水町邦彦. 化学教育台(日),1978,26:139.
[14] Popiel W J. Laboratory manual of Physical Chemistry, 1964
[15] 北京大学化学系物理化学教研室. 物理化学实验. 第3版. 北京:北京大学出版社,1997.
[16] 胡英主编. 物理化学. 第4版. 北京:高等教育出版社,1999.
[17] 黄寿祥,蒋萍初,许海涵. 果汁组成对BZ化学振荡的影响. 上海师范大学学报(自然科学版),1998,17(3).
[18] 李和兴,倪丽华等. 木糖-BZ体系化学振荡反应的研究. 上海师范大学学报(自然科学版),1996,25(3).
[19] 陶长元,颜红梅等. 酸度对BZ振荡反应的影响. 物理化学学报,2000,16(9).
[20] 李宗孝,原春兰. 以氨基酸为底物的B-Z振荡反应研究. 化学通报,1993(1).
[21] 贺占博. 设计振荡反应的简易方法——分解-耦合法. 化学工业与工程,1955,12(2).